Elements of
I[illegible]
Robotics

Barry Leatham-Jones
Head of Department of Engineering
Worcester Technical College

Pitman

PITMAN PUBLISHING
128 Long Acre, London WC2E 9AN

First published in Great Britain 1987

British Library Cataloguing in Publication Data

Leatham-Jones, Barry
Elements of industrial robotics.
1. Robots, Industrial
I. Title
629.8'92 TS191.8

ISBN 0-273-02592-9

ISBN 0 273 02592 9

Printed at The Bath Press, Avon

The author and publisher would like to extend sincere thanks to the following organisations who have contributed to the production of this book.

Aids Data Systems Ltd.
Asea Robotics Ltd.
Centa Transmissions Ltd.
Cincinatti Milacron Ltd.
DSR Systems Ltd.
Engineering and Scientific Equipment Ltd. (ESE)
Epson U.K. Ltd.
Harmonic Drives Ltd.
The Machine Tool Trades Association. (MTTA)
PGM Ballscrews Ltd.
RS Components Ltd.
Unimation (Europe) Ltd.

Contents

Preface

Central to the success of modern, flexible production and computer integrated manufacturing systems are four underlying themes:

- flexibility
- predictable and consistent output
- rapid response
- reduction in the hidden costs associated with manual production systems

In the realisation of these, industrial robots are set to make an important impact across many diverse manufacturing industries and disciplines. Second only to the computer, the scope and potential contribution offered by industrial robots in the manufacturing environment is probably greater than any other single form of advanced manufacturing technology. This view is supported by a sustained worldwide growth in robotic applications.

Inevitably, the demand for skilled robotic system engineers must be met and dedicated college courses are now being formulated in response to this demand. The subject of robotics draws together many, already established, engineering disciplines alongside many emerging skills. This book has been written to offer a comprehensive introduction to the technology, selection and application of industrial robots in the production manufacturing environment. It is anticipated that readers will be:

- Students of engineering being introduced to robotics for the first time through vocational or short college-based courses.
- Established engineers wishing to update their knowledge of industrial robots.
- Busy engineering managers who wish to gain an insight into industrial robots and their application.

The text covers the syllabus content of established courses in robotics at both advanced craft and introductory technician levels. In support of these study areas, related questions appear at the end of each chapter and a number of case study assignments are included.

In writing this book I am indebted to the various people and organisations that have responded so willingly to my requests for photographs and other material to appear within the text.

The book was produced using a BBC microcomputer and the View word processor. I should like to express my sincere thanks to Epson U.K. Ltd. for supporting the project by so generously providing an Epson RX-100 printer on which the manuscript was prepared.

Last, but by no means least, I must thank my wife Helen and my daughter Laura for their continued patience and caring support during the preparation of the book.

For Helen . . . who cares more than anyone

S. B. Leatham-Jones
Hoghton 1986

The Scope of Industrial Robots 1

1.1 Definition of an industrial robot

1.1/0 Automation and robotics

Robots are increasingly becoming a key feature of modern industrial manufacturing systems. They know few boundaries and can be applied across a broad spectrum of manufacturing disciplines in diverse industries.

Automation, the forerunner of modern robotic devices, is similarly employed across as wide an industrial spectrum if not wider. So what is a robot, and how does it differ from traditional automation?

The term **automation** can be simply described as the capability to operate without direct human intervention. Automatic devices have been on the industrial scene for more than 100 years. They are largely mechanical devices purpose-designed and built to perform a specific dedicated task. Because the configuration of such devices cannot easily be changed, a modern term labels this as **hard automation**. If the task or the component changes then this hard automation becomes redundant, or has to undergo physical modification to adapt its operation to suit the new conditions, or may be cannibalised and the component parts used again on other applications.

Hard automation is characterised by the following features:

a) It is relatively cheap and simple to implement, although it may be expensive and somewhat time-consuming to develop.
b) It is simple, efficient and reliable in operation since all motions are fixed and constrained within the structure of the device.
c) It is simple to control since the task is dedicated and there are no redundant movements or degrees of freedom.
d) It is relatively easy to safeguard in terms of human safety.
e) Because of its limited scope and inflexibility it may quickly become obsolete if the production task, component or product model changes.
f) Because of its inherent purpose-built design it requires little maintenance or human intervention.

Industrial robots also operate without direct human intervention and as such they form a sub-class of automation. In addition they exhibit certain other characteristics that set them apart from dedicated automatic devices.

1 *Industrial robots are computer controlled*
Most modern industrial robots are controlled by digital computer control systems. Computer control is necessary for a number of reasons:

a) To carry out complex mathematical calculations and coordinate transformations to manipulate the end of the robot arm in three-dimensional space.

b) To coordinate multiple axis movements in order to achieve smooth, accurate and repeatable control of straight lines and complex contours.

c) To simplify the design, development, programming and control of industrial robots.

d) To receive and process multiple inputs from a number and variety of external sensors to effect conditional operation of the robot.

e) To enable quick, responsive and reliable robot operation.

f) To enable relatively simple integration with other aspects of a computerised manufacturing environment.

2 *Industrial robots are re-programmable*
The movements and actions of an industrial robot can be determined initially by a human operator. A sequence of movements can be stored within a control system computer and the robot can be commanded to perform these movements repeatedly and with great accuracy. If the task, component or product model changes, the robot can be taught a new sequence of movements, thus easily adapting to the new conditions. The original program of movements need not be lost since it can be saved for subsequent modification or re-use at a later date.

Since robot operation is largely under **software control** (i.e. the control of a computer program), robots can be thought of as **soft automation**.

3 *Industrial robots have multiple movements*
Industrial robots can perform simultaneous, free-ranging three-dimensional manoeuvres in space, outside their immediate structural form. Six *degrees of freedom* of manipulation are provided on most industrial robots. This makes them adaptable to a wide variety of industrial tasks. Additional degrees of movement may be provided to increase the accessibility of some robots for certain applications.

4 *Industrial robots have interchangeable grippers*
Most industrial tasks involve transporting or manipulating tools, components, appliances or applicators to perform various manufacturing activities. The majority of these tasks involve the holding or gripping of items of varying size and shape. It is almost impossible to design a practical, universal holding device. In order to retain the flexibility of robot applications, most robots can be fitted with a variety of different 'hands' or *grippers*. These are known collectively as *end effectors*, since in many applications no actual holding or gripping takes place. For example, many robots have special tools such as welding heads, grinders, applicator nozzles, etc. permanently attached to the end of the robot arm, which do not hold or grip at all.

End effectors may be mechanical devices, electro-magnetic, pneumatic (vacuum cups) or a combination of these, by design. In certain cases they may consist simply of hooks, ladles or holders. It is becoming more common for robots to be applied to more than one task simultaneously, often necessitating a different end effector from one task to another. Under such circumstances an automatic end effector change may be programmed within the operational cycle. Robots are also capable of carrying out an end effector change without human intervention.

On the basis of the above characteristics an industrial robot may be defined as:

> *A computer-controlled, re-programmable mechanical manipulator with several degrees of freedom capable of being programmed to carry out one or more industrial tasks.*

In comparison with hard automation robots have many redundant features and links which can cause their operation to be slow, unwieldy and inefficient. Because of the many degrees of movement, precise control of robots is often difficult to achieve. The robots themselves are accompanied by complex and expensive computer-based control systems. The very flexibility that enables robots to perform versatile multi-directional movements also means that they are extremely difficult to safeguard to maintain human safety. So why are robots becoming so increasingly popular and displacing the widespread application of hard automation devices? Some of the reasons are as follows:

a) Robots offer immense flexibility since they can be applied to many different tasks and can be re-programmed in the event of a sequence or operation change. Different programs can also be saved to a backing store (magnetic tape or disc) for use at a later date. With a typical product life cycle of between 1 and 3 years such flexibility is highly desirable in terms of a sound capital investment, the time required to respond and adapt to changing conditions, and reducing changeover and re-tooling costs for new models or components. Obsolescence is effectively designed out. The addition of some form of locomotion to the robot enables either dynamic operation to be performed on moving targets, or the robot to carry out operations at positions outside the reach of its basic configuration.

b) Robots can handle a number of different cargoes. Different shapes, sizes, materials and forms of packaging can all be handled with the same comparative ease. Their programmability and the provision of sensory feedback enable them to deal sympathetically with cargoes that may be fragile or delicate.

c) Robots of various design configurations, shape, size and capability are available 'off-the-shelf'. Quality, accuracy, repeatability and reliability are available without the need to undertake time-consuming and costly design, development, building and testing programs required for hard automation devices.

d) The unit costs of modern industrial robots are very competitive and they may offer a more attractive proposition on a cost/performance measure than traditional hard automation, given their flexibility.

e) Since robots are controlled by software and often have motions that compare with those of the human body, they can easily be integrated into existing manufacturing systems with the minimum of workplace re-configuration.

f) Robots are compatible with, and can easily be integrated with, other computer-based manufacturing technologies. Their ability to carry out different actions conditional upon sensory feedback from the workplace enables them to work unattended by human operators. They offer the vital link in achieving unmanned and flexible computer aided manufacture.

1.1/1 Brief history of robots

The term 'robot' is derived from the Czech word 'robota' which means forced labour or slave. It gained popularity from 1923 following a play written by a Czech playwright called Karel Kapek. The play (called *Rossums Universal Robots*) ends by perfect human-like robots exterminating all human life, after becoming intolerant of their human imperfections. It is interesting that the human mistrust of robots as destroyers of jobs was farsightedly portrayed in Kapek's play. Indeed, the earliest commercial industrial robots were termed 'Universal Transfer Devices' by their manufacturers because of this very implication.

Robotics is really the combination of many already-mature engineering disciplines. Mechanical, electrical, pneumatic, hydraulic, electronic and, more recently, computer technologies all come together to form a functional robotic system. In addition, the discipline of design has to be harnessed when originating special-purpose end effectors.

Early practical robots were demonstrated in 1955, and the first commercial industrial robot was introduced in 1961 (in a die-casting application). Early robots were primarily controlled by electrical analog devices since the digital computers of this period were slow, bulky and expensive. Physical control of the various motions was accomplished by applying stepped voltage levels, set by potentiometers, to position servo-amplifiers. The different voltage levels represented different end positions of the arm. The arm positions were sequenced (programmed) via rotary relays of a kind used in early telephone systems. Limitations of these systems included:

a) Only a limited number of path movements could be programmed.
b) Movement was essentially between points and even relatively simple contours could not be traversed.
c) Relative motion (an incremental move from a previous position) was almost impossible to achieve.
d) Conditional movement (the capability to perform different movement patterns depending on external conditions) was virtually impossible. Robot operation was limited to simple predefined sequences even though the sequences could be re-specified.

Successful modern robots owe their flexibility and popularity to two important factors: their programmability and the capability of conditional operation as a result of sensory inputs from the workplace. Both factors have been realised by the rapid development of microelectronics throughout the 1970s. Such developments have brought fast, reliable, low-cost and compact digital computing power, in the form of the microprocessor and microcomputer, to many industrial applications. Indeed, the whole of manufacturing industry is being re-shaped by parallel developments based on digital computing technology. An obvious example is the widespread application of CNC (*computer numerically controlled*) machine tools.

Science fiction has long since demonstrated its remarkable ability to predict (and often influence) eventual science fact. Robots continue to command a high standing in science fiction and maybe this is indicative of the role they will continue to assume in the future. As long ago as 1942 the prominent science fiction writer Isaac Asimov formulated three '*Laws of Robotics*'. They are quoted here in recognition of the research in the field of *artificial intelligence*. The ultimate conclusion will be the development of a robot with the capability to make strategic decisions and act on them accordingly. Such decisions will be made with cool precision and without due regard to human attributes such as common sense, judgement, emotion, fear, love, rationality, etc.

The three Laws of Robotics, as propounded by Asimov are:

1 A robot may not injure a human being or, through inaction, allow a human being to come to harm.
2 A robot must obey the orders given to it by human beings except where such orders would conflict with the first law.
3 A robot must protect its own existence as long as such protection does not conflict with the first or second laws.

The development of a thinking robot may eventually prove to be the easy part of the exercise. It remains to be seen whether the moral and ethical questions and problems can be similarly overcome.

1.1/2 Classification of industrial robots

There are a number of ways in which industrial robots may be classified, some more useful and more meaningful than others. Such classifications are merely broad statements describing certain important features possessed by the robots. A classification should not be confused with a specification. A **classification** is a means of placing a particular robot into a broad category or group whereby it can then be compared with like robots in the same group, as to its suitability for a particular application. A **specification** is a factual description of the robot's size, capacity, capability, geometry and so on. It is possible for a single robot to fall within many systems of classification depending on the purpose for which it is being considered. Some common means of robot classification are as follows:

Configuration Industrial robots may be grouped according to their physical design or geometrical structure. This is known as their configuration. For example, some robots have arms that rotate whilst others may be constrained to move only in straight lines. The terms 'degrees of freedom' and 'degrees of movement' become important when considering robot configurations, since they can significantly influence the space envelope within which the robot can operate. These terms and a description of the common industrial robot configurations are explained in Chapter 2.
Control System Important considerations when selecting a robot concern its positioning accuracy, its repeatability and its capability to traverse smooth and often complex contours. Such considerations are a function of the type of control system employed. Terms like 'servo and non-servo controlled', 'open or closed loop control' and 'point-to-point or continuous path control' describe common control systems classifications. These terms and control systems in general are discussed in Chapter 3.

Power Transmission Industrial robots may be powered by hydraulic, pneumatic, electrical and mechanical devices. There may be conditions imposed by an application which will influence the power transmission required. For example, high power lifting applications dictate that a hydraulically powered robot needs to be specified. In many cases robots employ more than one power system for various aspects of their operation. They are termed *hybrid systems*. Power transmission and axis drive systems are discussed in Chapter 3.
Programming System Robot sequences may be 'programmed' in a number of ways. These range from the physical positioning of limit switches to determine a fixed sequence of movements, to writing programs in specialist robot programming languages, away from the robot. The programs can then be loaded into the robot control system memory at a later time. Programming considerations are discussed in Chapter 7.
Generation Industrial robots may be considered 'dumb', 'clever' or 'intelligent' dependent upon the generation to which they belong. How these terms are defined and the different types of generation are discussed in Section 1.1/3.
Application Industrial robots may be classified according to the particular task for which they are employed. For example, a supplier may concentrate on 'welding robots' or 'assembly robots' or 'material handling robots', etc. Although this method of classification may not appear to be that useful, it at least alerts a supplier or user to the application area that needs to be addressed. This may then suggest a further refinement of classification in selecting the correct robot for the job. It may also be relevant to those robot manufacturers or suppliers that provide turnkey systems. A *turnkey system* is an application where a sole supplier takes responsibility for design, specification, supply, commissioning and support for a total project. This may encompass the design of premises, workplace layouts and other essential support equipment. Industrial applications of robots are discussed in Section 1.3.

Although most robots exhibit common characteristics they vary in complexity and capability. This is equally true within the various classification groups. In general, the more flexible, adaptable and highly specified the robot, the more complex and expensive it becomes.

1.1/3 Robot generations

Robot technology, like electronic technology, is developing in distinct phases or *generations*. At present three generations can be identified although the third is still largely at the research stage.

First-generation robots
First-generation robots can be likened to devices that operate according to a strict, fixed sequence of events. They are known as 'dumb' robots since they faithfully reproduce the programmed sequence whether work is present or not. They cannot detect any change (i.e. the presence or absence of a component) in the surrounding environment, and cannot therefore modify their actions accordingly. Although they are programmable, programming is done by altering the physical positions of limit switches, re-setting stops on indexable drums, or replacing cams to alter movements. The successful use of first-

generation robots depends, almost exclusively, on ensuring that the correct components are presented to the robot in the correct place, in the correct orientation at the correct time. This usually entails the provision of support equipment (e.g. conveyors or feeders).

Second-generation robots

Second-generation robots can be classed as 'clever' robots. They are equipped with a range of sensors, and the necessary computing power, to modify their actions in response to small detectable changes in the surrounding environment. For example, proximity sensors can differentiate between a number of different components and 'instruct' the robot to execute a different sequence of events depending on the component which has been identified—for instance, large components being transferred from a conveyor and placed into one bin, and small components being picked from the same conveyor and placed into a second bin. Second-generation robots (or more correctly their control systems) are necessarily more complex, and therefore more expensive, since they have to be provided with a range of sensors and the associated control software. Latter developments enable such robots to be linked to host computers and have access to CAD databases.

Third-generation robots

Third-generation robots are the 'intelligent' robots. They are at present only in the research stage. They will be characterised by their ability to plan, make strategic decisions and execute tasks 'intelligently'. They are likely to be programmed to maximise (or minimise) some defined objective. Development of third-generation robots will depend to a large extent on parallel developments being made in artificial intelligence (AI) software systems.

First-generation robots, whilst still in use, are being gradually superseded by second-generation robots. Second-generation robots are emerging in larger numbers and are finding greater application in all industrial tasks. They are, however, under continual development and are destined to find even wider application. It should not necessarily be assumed that the ultimate industrial vision is one of third-generation robots assuming dominance in the manufacturing environment. The industrial application will continue to determine the means of achieving the desired goal. It may be entirely appropriate in those applications that do not necessarily require the flexibility of robots, to continue to apply hard automation techniques. In applications where small and medium batch quantities are the norm, readily programmable first- or second-generation robots may continue to be entirely suitable.

1.2 The need for industrial robots

1.2/0 Competitive benefits

Robots are usually expensive items of capital equipment. Their adoption also implies significant costs in maintaining support activities such as programming, maintenance and production engineering functions. The utilisation of robots within a manufacturing environment therefore has to be justified against alternative means of accomplishing the same task or tasks. In management terms they have to be 'cost effective'. In essence their adoption must contribute to an overall gain in efficiency. This means either an increase in productivity or an attendant reduction in production costs. This in turn translates, eventually, into assuming a more competitive position in the marketplace or contributing to an increase in net profitability. Cost effectiveness may be measured in a number of ways and by a number of criteria. A number of possible criteria are suggested below:

1 *Uniformity of output*

If uniformity of output can be achieved, then benefits accrue in terms of reduced levels of inspection and quality control, streamlined fitting and assembly procedures, reduced costs of scrappage and/or re-work items, smoother production, enhanced customer satisfaction, high corporate reputation, etc.

2 *Accuracy, repeatability and consistency*

These qualities imply many of the benefits already mentioned. In addition it is possible to broaden the scope of work that can be carried out (within the specifications offered) and significantly improve cycle times in tasks that require high levels of human involvement. More importantly, perhaps, is that consistency of output can be maintained over the complete working period and will not deteriorate through fatigue, boredom, carelessness, etc.

3 *Predictable cycle times*

Efficient production control relies on sound planning, which in turn relies on predictable operation and throughput times. For many reasons predictable cycle and operation times often turn out to be elusive in a manufacturing environment. Automation allows cycle times to become predictable, which allows production schedules and delivery dates to be accurately planned and, above all, met. Additionally, cost estimates become similarly predictable allowing tighter financial control and competitive tendering.

4 *Reliability of operation*

Achievement of target levels relies not on achieving the fastest cycle times but on achieving reliable planned output. Downtime, for whatever reason, must be minimised and, if at all possible, eliminated. Human workers, for example, are perhaps more susceptible to contributing to downtime than automated devices. Reasons include accident, injury, illness, tiredness, boredom, carelessness, lateness, fatigue, toileting, washing and feeding, absence, industrial action and so on. Robots and automated devices remain free from the majority of these human 'failings'.

5 *Availability for work*

The efficient management of resources is a prerequisite to a flexible and responsive organisation. Ideal conditions suggest the employment of devices that can be put to work at any time, do not require frequent rest periods, can work at short notice and at unsocial times including holiday periods, can work unattended without constant supervision, can sustain unpleasant working environments, and are tolerant (and capable) of frequent change. All these do much to enhance the flexibility and responsiveness of an organisation. The costs of running robots remains roughly stable no matter what part of the working calendar is considered. The costs of employing human labour can rise considerably during overtime, unsocial or holiday periods.

6 *Flexibility and adaptability*

Flexibility and adaptability within an organisation are required to sustain its competitive operation in the light of changing circumstances. Such circumstances include the emergence or strengthening of external competition, trends in fashion and taste, seasonal influences, product life, the emergence of new techniques and so on. There must therefore be an inclination to invest in resources (both physical and human) that are capable of such flexibility and adaptability. Moreover, robots are programmable devices. Tasks and sequences, once programmed, can be retained for recall and subsequent use at a later date. Changeover times between tasks are thus minimised. The investment in programming and production engineering development work can be realised many times over.

7 *Running costs and overheads*

Runnings costs may contribute significantly to overheads. *Overheads* is the term given to describe those costs which have to be met irrespective of output. It follows that, for greatest efficiency, either production has to be maximised for a given level of overheads or, since production cannot be infinite, overheads have to be reduced to as low a level as possible. Overhead items associated with servicing robot installations include programming, maintenance, production engineering development, depreciation and so on. Overheads associated with employing human operatives include legal contributions and services, welfare and safety services, supervisory services, timekeeping, work study and administrative services, environmental services and so on. Whole departments often have to be engaged in such activities to support the 'production' workforce.

8 *Reductions in 'hidden' costs*

The term *hidden costs* means those costs that can significantly affect the overall costs of production, yet cannot be accurately determined. There are many examples. *Work-in-progress* (WIP) refers to stocks of part-finished items awaiting further processing. High work-in-progress means large amounts of working capital tied up in part-finished stocks and raw materials as well as consuming valuable storage space. Poor production control and badly organised production facilities can make it difficult to keep track of the whereabouts of components, or batches of components, on the shop floor. As a consequence, armies of *progress chasers* (sometimes called *expediters*) are often employed to establish the exact whereabouts of components. Inaccurate costing leading to inordinately high cost estimates may lead to the loss of orders and consequent loss of production. The true effects and costs of the above are difficult to determine. They are direct consequences of the inability to control certain elements of the organisation. Measures that enable control of these elements to be returned to management should contribute to an increase in efficiency and/or profitability.

9 *Integration with other aspects of the organisation*

The automatic operation of isolated parts of a system inevitably leads to the notion that the parts themselves can be linked together to form an automated production system. Such ideas are indeed reality. *Flexible manufacturing systems, CAD/CAM systems, computer integrated manufacturing systems* are all modern concepts of fully automated production facilities. Inevitably, robots play a significant role in each of them since they offer the possibility of simple and effective integration.

Hard automation (as defined in Section 1.1/0) and human operators represent the alternative means against which the introduction of robots will be compared and justified. A comparison of these three possibilities, based on the above criteria, is tabulated in Fig. 1/1. The comparison can be made a little more graphic by applying simple score values to the attributes allocated within the table. The results of carrying out this exercise are shown at the foot of the table.

Fig. 1/1 Simple comparison of hard automation, human operators and robots

	Hard Automation	*Human Operator*	*Robot*
(a) Uniformity of output	EXCELLENT	FAIR	EXCELLENT
(b) Accuracy, repeatability and consistency	EXCELLENT	FAIR	GOOD
(c) Consistently predictable cycle times	EXCELLENT	FAIR	EXCELLENT
(d) Reliability of operation	EXCELLENT	FAIR	GOOD
(e) Availability for work	GOOD	FAIR	EXCELLENT
(f) Flexibility and adaptability	POOR	EXCELLENT	GOOD
(g) Running costs and overheads	LOW	HIGH	MEDIUM
(h) Reduction of hidden costs	FAIR	POOR	GOOD
(i) Integration with automation	POOR	FAIR	EXCELLENT
(Score comparison)	(24)	(21)	(31)

SCORE	
EXCELLENT/HIGH	4
GOOD/MEDIUM	3
FAIR	2
POOR/LOW	1

People are still the most useful resource available to management. They can think, apply judgement in unforeseen circumstances, adapt to new situations, be creative and imaginative, and be capable of improving on the skills they possess. They also have shortcomings, many of which are related above. It makes sense, from a management point of view, to harness these qualities and realise the full potential of the human workforce. The introduction of robots to carry out menial, repetitive and unpleasant tasks can release valuable human talent for deployment elsewhere, for example in tasks that cannot be performed effectively by robots, or that require human intellect and/or judgement in their solution. In addition, time lost through physical disorders, accelerated by poor or arduous working conditions, can be dramatically reduced. Robots are stronger and have greater reach than human operators. Additional effective man hours can thus be made available, together with the reduced cost of supporting the fewer workers laid off through industrial injury. These, and other, hidden benefits should also be considered as legitimate reasons for introducing robots.

1.2/1 Drawbacks of robotisation

The drawbacks of introducing robots into an already established organisation are much the same as introducing any high-technology equipment.

The first drawback is one of capital investment in high-cost equipment. Robots, unlike most other computer-based equipment, are increasing in price. (Costs concerning robot installations are discussed in Chapter 9.)

The economics of automated production are sensitive to machine reliability. Continuous operation (especially under multi-shift working) will quickly exploit any deficiencies. Robots are high-technology units and as such require maintenance of a different nature to that of hard automation or conventional machinery. The establishment of a resident, high-calibre maintenance facility is a prerequisite of adopting robots. Such a maintenance facility needs to be multi-disciplined in order to support the different technologies embodied within modern robot and control system designs.

If robots are not to be idle or inefficient in operation, for long periods of time, planned support facilities are essential. Programming, production engineering, tooling (end effector) design are three obvious areas that require support.

Robots will, to many, be a new and bewildering discipline. The existing workforce may need training or re-training. New skills may have to be introduced into the organisation from outside. A managed system of training will need to be very carefully considered.

The introduction of robots will inevitably lead to the replacement of existing jobs. Whilst many new jobs may be created in the various support functions, the demise of many jobs will create surplus labour that needs shedding or redeploying. Planned alternatives for those members of the workforce so affected must be carefully considered.

The introduction of just one robot into an organisation is a major undertaking in both technological and human terms. Quite apart from the selection of a suitable robot configuration and control system, it is essential that careful pre-planning and feasibility studies are conducted. In addition, full consultation with the workforce will be amply repaid in ensuring a smooth transition and a ready acceptance.

1.3 Industrial applications of robots

1.3/0 Component handling

Component handling is an ideal application for an industrial robot. It is a repetitive operation, carried out often under unpleasant and arduous working conditions, which requires little skill. Component handling tasks usually involve fairly simple manoeuvres, with only modest accuracy being required. End effector designs can usually be kept simple and functional, consistent with the shape, weight and material of the components involved. Component handling tasks assume that the robot in some way grips or holds the components concerned.

Component handling applications vary widely:

a) Loading and unloading conveyors.
b) Loading and unloading machine tools (components).
c) Loading and unloading machine tools (cutting tools, etc.).

d) Machine-to-machine transfer.
e) Unloading die-casting machines.
f) Unloading injection moulding machines.
g) Loading and unloading jigs and fixtures.
h) Loading and unloading presses.
i) Extracting, transporting and transferring castings
j) Loading and unloading furnaces.
k) Component handling for forging and stamping processes.
l) Component transfer for heat treatment processes.
m) Component handling for plating and coating processes.
n) Producing investment casting moulds.
o) Handling dangerous or toxic materials.

Many of the above applications may also include associated ancillary operations such as applying lubricants, air blast cleaning or the positioning of cores or inserts. These operations can be perfectly synchronised within efficient cycle times and during high-volume, repetitive production runs. Safety can be designed in using sequential interlocks whereby subsequent operations or movements cannot be initiated until previous conditions have been fulfilled.

The application of robots to some of the above tasks may not always be quite straightforward. There are many practical production engineering problems that need to be addressed. For example, consider using a robot to load and unload a CNC turning centre. The component being unloaded may be of considerably different size and shape to that of the raw material from which it is produced (and loaded). The end effector design must be able to accommodate these factors. Consider also using a robot to load components into jigs or fixtures. Almost by implication, the use of jigs or fixtures suggests that the components are of irregular shape. If this is the case, then quite obviously they must be loaded the right way up and in the correct orientation. It is quite difficult (although not impossible) to get the robot to identify the orientation of components presented to it at random. Some means of feeding and orientating the components before they reach the robot may well have to be devised.

1.3/1 Palletising and packaging

Many component handling applications involve the packaging or palletising of finished products. *Palletising* is the term used to describe the ordering and placement of products or components onto pallets following a specified arrangement pattern. Boxes pack more efficiently in some arrangements than in others. This is termed *nesting*. Bags and sacks of granular or powdered material will be more stable in certain arrangements than if they are simply stacked on top of each other. These examples are illustrated in Fig. 1/2. Such considerations may be important, since they can considerably affect factors such as shipping costs and safety during transit.

Packaging operations also often involve the placing of spacers and other packaging materials in addition to the components themselves. Sorting, packaging, handling and labelling errors can be virtually eliminated. Characteristics of palletising/de-palletising and packaging/unpackaging operations often require:

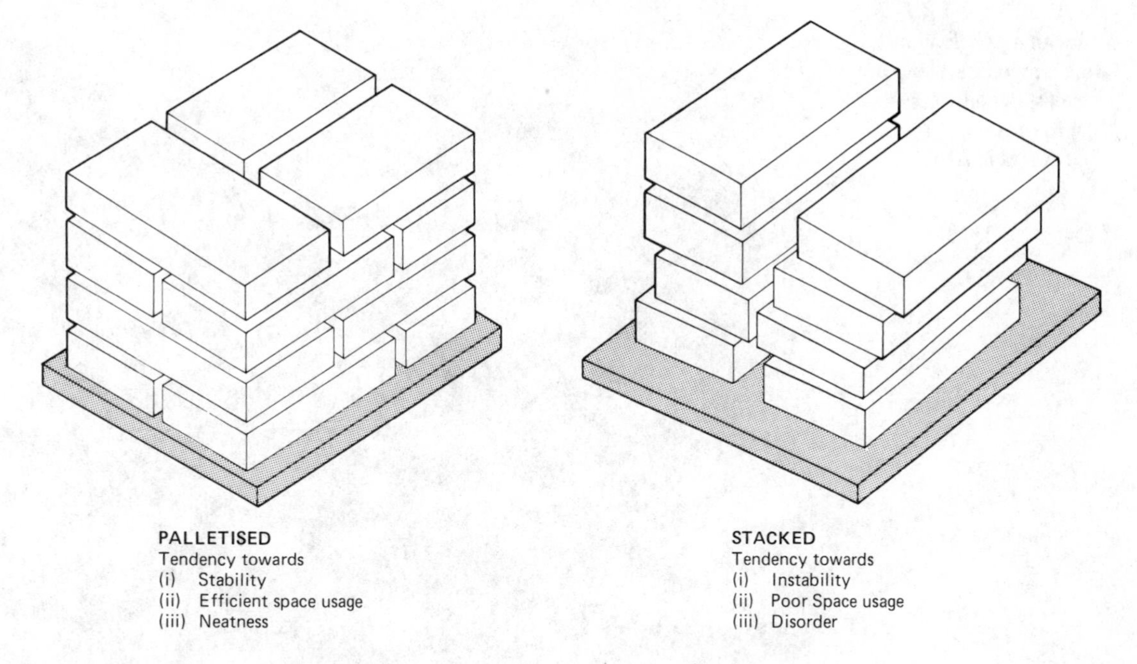

Fig. 1/2 Palletised products may result in more stable arrangements

a) Moderate accuracy.
b) Simple geometry and control.
c) High load-carrying capacity.
d) Good reach and mobility.
e) Sympathy with delicate cargoes.
f) Variety of gripping methods.

1.3/2 Assembly

Assembly operations are operations that mate or position two or more components relative to each other, so that they make a single entity. It excludes the operations of actual joining by, say, welding, screwing, bolting or glueing. These operations will be discussed in following sections.

Assembly operations by robots are forming an increasingly large proportion of robot applications and are continually being researched. Assembly tasks appear to be ideal applications for robots. To human operators they are repetitive and monotonous operations often requiring much practice to perfect and offering little physical movement or exercise. Assembly environments are often noisy and uncomfortable. Payment is often made on a 'piecework' basis which imposes stress and fatigue on the assembly worker. Fatigue and boredom can often lead to unreliability, which may also present safety hazards.

However, assembly tasks require high levels of concentration, high manual dexterity, close hand/eye coordination and a developed sense of 'feel'. Human workers also have the ability to apply quality checks on the component parts of the assembly, detecting flaws and rejecting items that may not mate properly. They can also pick randomly arranged components from storage

Machine tool loading and unloading using a revolute robot configuration
[*DSR Systems*]

Component placements and sub-assembly operations
[*DSR Systems*]

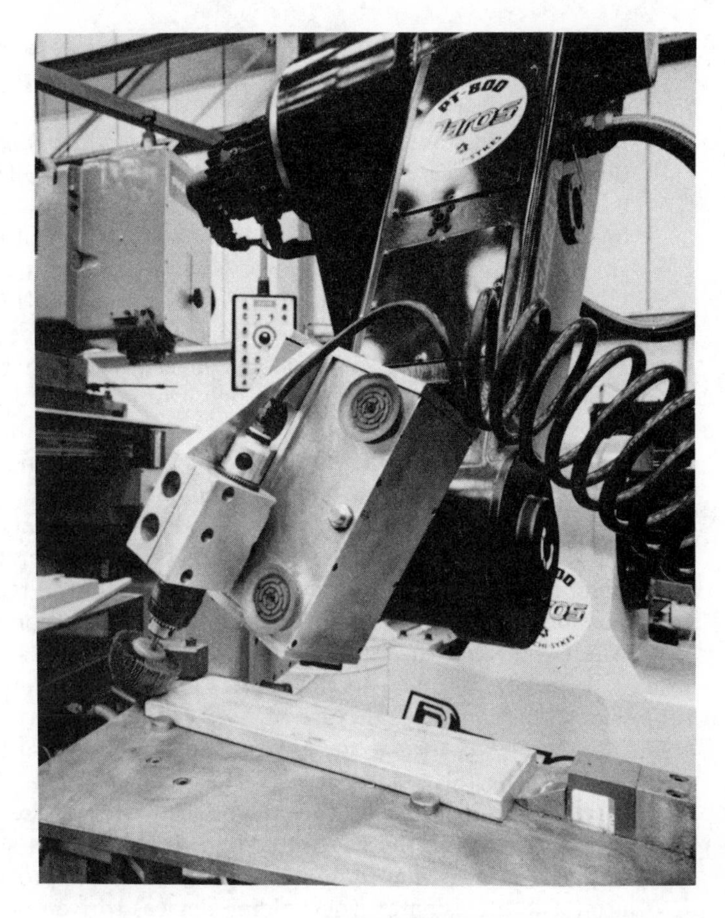

Loading, unloading and de-burring operations utilising a multi-functional end effector [*DSR Systems*]

containers with ease and speed. These 'human' attributes prove difficult to implement in present robotic devices and are the subject of much continuing research. Successful implementaton relies on sophisticated sensing devices, which indicates the use of second-generation robots.

Assembly tasks largely comprise movements in a horizontal plane for positioning, coupled with vertical movement for insertion or assembly operations. Such movements, although relatively simple to implement, take longer to perform by mechanical means than by human means. This is because distances traversed by robot arms, to achieve positioning and assembly, are usually considerably longer than those traversed by humans, as a result of constraints imposed by the mechanical design of robot configurations.

In the motor vehicle industry, robots are applied in assembling electrical wiring looms. Locating pins are first arranged plug-board fashion (by the robot) to allow the electrical cables forming the harness to be threaded between them according to a pre-determined route. The robot then selects pre-cut lengths of the various gauge cables and proceedes to lay them out to form the wiring loom. Different vehicles require different loom designs and these can, once programmed, be retained on a backing store device to be recalled as required.

1.3/3 Welding and cutting

Welding tasks, in particular spot welding, represent one of the largest application areas for industrial robotic devices. *Spot welding* is a form of electrical resistance welding whereby sheet metal is joined by a number of individual resistance welds that form characteristic 'spots' due to localised heating. Motor car bodies are fabricated in this way. Each body panel requires a large number of uniform high-quality welds, accurately and reliably placed in the correct position. Simple point-to-point control is all that is required, since welding always takes place at the end point of a particular move. Generally, these robots need to have a large payload capacity, because electrode heads and welding transformers can be large, heavy and unwieldy. Robots make an excellent economic investment in spot welding applications. Welding speeds well in excess of one spot per second can be achieved.

Developments, using more complex continuous path control systems, now allow three-dimensional fusion welding processes to be accomplished. *Fusion welding* describes those welding techniques whereby the parent material, of the parts to be joined, actually melts and fuses together. Examples are electric arc welding, MIG (metal inert gas), TIG (tungsten inert gas) and CO_2 welding. Filler metal is usually continuously fed by motorised reels or drums. It is also possible, using second-generation robots, to have 'joint following' capability where the robot can track a particular joint path or seam.

All robotic welding applications require some means of *parts presentation*. This usually involves the use of fixtures or power-driven manipulators. It is often easier to arrange for a fixture to be rotated, at a constant speed, and to program the robot to traverse a linear path. This allows complex three-dimensional welds to be produced whilst lessening the demands on programming and the control system. It also allows access to the complete periphery of a circular component where this might prove difficult for a jointed arm. The provision of workholding devices (driven or not) often represents a significant proportion of the cost of a robotic welding installation. Component parts must be correct to drawing before they are presented for welding. There is little scope, during robotic welding, for 'making parts fit' by applying *tack welds* to correct any misalignment on assembly, or misshape during manufacture.

By replacing the welding head with a cutting torch it is also possible to achieve *flame cutting* under robot control. However, a more common technique finding increasing application is *water jet cutting*. Relatively soft materials such as rubber, plastics, moulded foam, leather, cardboard, etc. may be cut or trimmed by this method. A high-pressure water jet (operating at around 350 MPa or 50 000 psi) is forced out of a nozzle of diameter 0.5 mm and collected by an output nozzle. The jet of water exposed, between the nozzles, forms a highly clean and efficient cutting edge. Abrasive particles can also be introduced into the water jet for cutting thin metals. The robot may either hold and transport the water jet itself around a fixed component or manipulate the component about a fixed jet arrangement. Either mode enables close and intricate shapes to be cut quickly and accurately.

Spot welding of motor car bodies using an articulated robot configuration
[*ASEA Robotics*]

Gas shielded arc welding on components mounted on a rotary manipulator
[*DSR Systems*]

1.3/4 Painting and coating

Another large industrial application for robots is *paint spraying*, particularly of components continually fed on overhead conveyors. The programmability of robots is ideally suited since components do not necessarily have to be fed through in any particular order. Different components can be sensed and identified and different spraying sequences called up from a number of pre-prepared move sequence programs. It also replaces the human operator in what is an unpleasant and unhealthy working environment.

The approach adopted by most robotic spray painting applications is to mimic the dexterous movements of a skilled and experienced paint sprayer. This is an isolated case where a machine is constrained to work in the way a human works. Most spray painting tasks involve traversing complex contours and therefore a robot incorporating continuous path control and many degrees of movement is essential. Efficiency and economy above that of human application can easily be achieved using robots. A more uniform surface coating, fewer rejects and a paint saving of more than 30% (through less wastage due to overspray, beyond surface edges) have been quantified. Air volumes can be considerably reduced both for the spraying operation itself and for environmental air changes. This translates directly into lower energy costs. Robots can consistently achieve accuracy, repeatability and increased speed, when used for decorative painting of trim lines and so on. First-generation robots (without sensory feedback) are adequate for most painting applications.

Other similar coating processes undertaken by robots include underseal application to the underside of motor bodies, the application of glazes to ceramic ware, enamel spraying, chopped glass fibre and resin spraying, and water jet cutting and cleaning operations.

1.3/5 Fettling and cleaning

Fettling is a process commonly associated with the cleaning up of castings. When sand cast components are extracted from their moulds after casting, they have two redundant masses of cast metal joined to them. These are known as the *runner* and the *riser* and are the means by which the mould is poured and filled. Similarly, on die-cast or injection moulded components the attached *sprue* must be removed. Investment castings usually consist of a number of components radiating out from a central tree structure. The individual components need separating and the central spine is then discarded. On forged components, where the forging dies meet, excess metal is often squeezed out of the sides of the die, forming what is termed a *flash*. All this unwanted (often waste) material has to be removed since:

a) It is surplus to requirements.
b) It renders the component unworkable for future operations or unsaleable because of its presence.
c) It may join two or more different components that need separating.
d) It may represent a considerable investment in raw material that may be reclaimable for future use.
e) It is not called for on the drawing and may not pass inspection or quality checks.

These operations may be accomplished by the robot gripping the components and applying them to various tools, or alternatively, robot-held tools may be applied to the components. It is of course possible for the robot to grip, transport and clamp the component into a pre-set position before picking up a fettling or cleaning tool to carry out the particular task. Such tools include flame cutters, bandsaws, pneumatic chipping hammers, grinders, polishers and abrasive belts. Such operations can be repetitive, dirty, noisy, unhealthy and sometimes dangerous tasks for human workers to undertake. The control system needs to be capable of describing often complex continuous path movements.

1.3/6 Applicating and dispensing

Applicating and *dispensing* operations involve the metering and delivery of measured quantities of consumable products in a uniform manner. The products can be liquid, powder, granular or viscous in nature. Common products include adhesives, mastics, sealants and caulking materials. Depending on the application, often quite complex paths have to be traversed. Uniform beads of adhesive, for example, may have to be applied to the lid of a shaped casing before it is bonded to its mating half. This requires high-accuracy high-repeatability contouring control, but relatively low payload capacity. Adhesive application generally requires close control in a single plane only, since the majority of parts to be bonded are flat. Many of these materials are volatile and toxic, and deploying robots can be justified on safety and environmental grounds.

1.3/7 Inspection and testing

Robots are increasingly used for inspection and testing. One creative example is *joint sniffing*. In the motor car industry, when windscreens are installed, they have to be tested for being leakproof. Instead of applying water to the sealed joint and visually testing for leaks, air is used. The inside of the car is first air pressurised at a relatively low pressure. A robot, equipped with a sensitive detector at the end of its arm, then traverses the external contour of the fitted windscreen. Any escaping air, signifying a leak in the seal, can be detected by the robot sensor and action taken accordingly.

Other applications involve dimensional checks. On smaller items the components can be picked up by the robot and applied to gauging systems. On larger components and structures, robots can be used to check dimensions by applying probes at predetermined points. The dimensions at these points will be known (from the original design drawings) and can be compared with the readings fed back via the probing operation.

1.3/8 Light machining

Robots may be used to transport portable cutting tools for performing light machining operations. Applications such as routing and repetitive drilling are common.

Routing machines, for example, can be heavy and cumbersome to handle

Valve seat finishing operations on internal combustion engine cylinder heads
[*DSR Systems*]

for human operators, quickly causing fatigue and a consequent loss of accuracy and repeatability. The routing process also produces large quantities of fine dust or flying particles which present a hazard.

The drilling of large quantities of small-diameter holes in awkward positions, such as in the manufacture of aircraft structures, may be accomplished by robots manipulating the drilling tool in conjunction with drill template guides. The robot can also be programmed to perform ancillary tasks such as drill-head changing and component handling as support functions to its primary task.

Questions 1

1 Outline the main differences between hard automation and industrial robots.
2 Explain why the application of industrial robots is often preferred to that of hard automation.
3 State *seven* engineering disciplines that find application in the design of industrial robotic systems.
4 What are the 'three laws of robotics', and where would they find relevance in modern industrial robots?
5 Briefly explain *six* ways of classifying industrial robots.
6 Explain, giving examples, what is meant by the term 'robot generation'.
7 Compare and contrast the use of industrial robots with that of hard automation and human operators.
8 Discuss the competitive advantages claimed for employing industrial robots in manufacturing industry.

9 List *four* reasons *for* and *four* reasons *against* industrial robots, from the point of view of the industrial manual worker.

10 List *four* reasons *for* and *four* reasons *against* industrial robots, from the point of view of management.

11 Briefly outline the possible drawbacks associated with employing industrial robots.

12 Explain the term 'palletising' and why it should be carried out.

13 State *ten* applications, involving component handling, to which industrial robots may be successfully applied. Give reasons for your choice of applications.

14 Suggest *three* possible problems that could be involved in applying industrial robots to component handling situations.

15 Discuss the advantages and limitations of industrial robots in assembly tasks.

16 Outline *two* likely pre-conditions imposed on components being considered for manufacture by robotic welding.

17 Briefly explain why industrial robots are extensively applied in painting and coating applications within manufacturing industry.

18 State, giving your reasons, the characteristics desirable in robot systems being applied to applicating and dispensing tasks.

19 Briefly explain the role of the industrial robot in inspection, testing and light machining operations.

20 When were the first robots used in industry, in what application, and why were they termed 'Universal Transfer Devices'?

Industrial Robot Configurations 2

2.1 Physical robot configurations

2.1/0 Robot selection

Although industrial robots take many different physical forms there is no absolute 'best' configuration. 'Best' will be determined by the application concerned. Different configurations for example, have different advantages which make them more suitable for certain tasks.

Physical and geometrical configurations, however, are not the only considerations in the choice of robot for a particular task. Others include:

a) Reach.
b) Working volume or envelope.
c) Payload capacity.
d) End effector capability.
e) Accuracy.
f) Repeatability.
g) Manoeuvreability.
h) Speed of operation.
i) Form of motion (point-to-point *vs.* continuous path).
j) Provision of sensory feedback.

It is important that the elements and requirements of the immediate task, together with any related, anticipated, or likely future developments, are fully explored before a final decision is reached on selection. A feasibility study should always be conducted, since the outcome could indicate that robotisation is inappropriate for a particular application. Such an outcome is entirely valid and helpful as it may prevent large capital expenditure on an ineffective resource item.

Industrial robots vary greatly in the number and types of movement they possess, their geometrical arrangement and their physical make-up. These factors together formulate the final design of the robot, called its **configuration**. The configuration of a robot determines to a large extent those tasks to which it is best suited. It affects such things as the robot's manoeuvreability, its payload capacity, its reach and so on.

A robot configuration typically comprises a number of rigid, jointed members that together form the **manipulator** or *arm*. These members are made from a material strong enough to withstand a variety of applied forces and loads. Considerations of strength and rigidity must take into account:

a) The payload (weight of the cargo) to be handled.
b) The weight of the end effector.
c) Dynamic and inertial forces imposed as a result of rapid accelerations and decelerations.
d) The weight of the robot structure.

Typical materials include steel, aluminium and high-strength composite materials. Since the emphasis is on high strength and low weight, most structures are fabricated from structurally efficient sections or are hollow. The latter have the added advantage that transmission and service elements (air lines, hydraulic hoses, electrical cables, etc.) may be safely and neatly contained within the structure itself. Robots constructed on a modular basis can greatly assist maintenance, as well as reduce the number of components that have to be stocked as spares.

Seven basic industrial robot design configurations may be identified.

2.1/1 Cartesian or rectangular configuration

The **cartesian configuration** provides for three linear axes of movement at right angles (sometimes termed *orthogonal*) to each other. The modes of movement are similar to those of a milling machine, providing movement in X, Y and Z axes. It may also be termed a *rectangular configuration* since its working range sweeps out a three-dimensional rectangular volume. Particular advantages of this configuration include:

a) Easily controlled/programmed movements.
b) High accuracy.
c) Accuracy, speed and payload capacity constant over entire working range.
d) Control system simplicity.
e) Familiar X, Y, Z coordinates easily understood.
f) Inherently stiff structure.
g) Large area coverage.
h) Large payload capacity.
i) Structural simplicity, offering good reliability.
j) Easy to expand, modular fashion.

This robot configuration finds application in those areas where linear movements and high accuracy are demanded—for example, manipulation of components through apertures (i.e. furnace doors, machine openings and similar confined spaces), or pick-and-place applications where the workplane is essentially flat. This particular configuration lends itself to **modularisation** in that it is relatively easy to bring about extensions of its movement by extending its linear axes. Two implementations of the cartesian configuration are illustrated in Fig. 2/1.

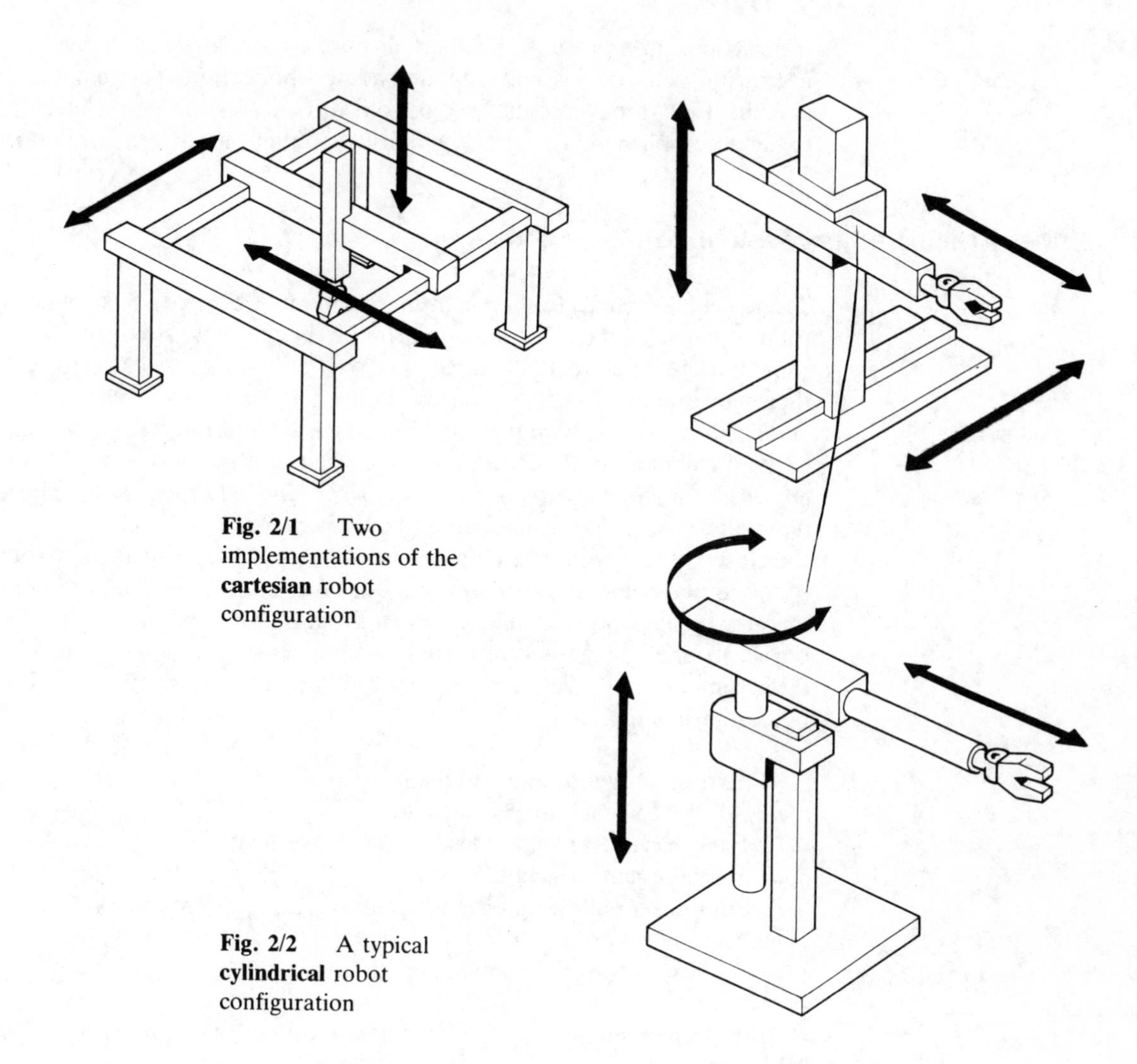

Fig. 2/1 Two implementations of the **cartesian** robot configuration

Fig. 2/2 A typical **cylindrical** robot configuration

2.1/2 Cylindrical configuration

The **cylindrical configuration** combines both vertical and horizontal linear movement, with rotary movement in the horizontal plane about the vertical axis. It is so called because its motions sweep out a partially cylindrical working volume. Particular advantages of this configuration include:

a) Easily controlled/programmed movements.
b) Control system simplicity.
c) Good accuracy.
d) Fast operation.
e) Good access to front and sides.
f) Structural simplicity, offering good reliability.

This robot configuration finds application in radial workplace layouts where the work is approached primarily in the horizontal plane and where no obstructions are present. Such applications include small 'circular' manufacturing cells or loading and unloading applications servicing conveyor systems. Linear movements through both vertical and horizontal apertures are easily accomplished. A typical cylindrical configuration is illustrated in Fig. 2/2.

2.1/3 Articulated or revolute configuration

The **articulated configuration** comprises a number of rigid arms connected by rotary joints. Rotary movement around the base is also provided. It is sometimes referred to as a *'jointed-arm' configuration* or, because it resembles the movements of the human body, it may also be termed an *'anthropomorphic' configuration*. The reach is determined by the length of each arm member in the configuration. Certain configurations may have either an *open kinematic structure* or a *closed kinematic structure*. These are illustrated in Fig. 2/3. The open structure is more flexible but less rigid than the closed structure. Since all movements are produced by angular rotations of the joints, complex calculations and transformations are required to move the arm through straight-line motions. Programming the arm (other than by physically moving it) is often troublesome, since it is not easy to visualise three-dimensional free-form movement. Particular advantages of this configuration include:

a) Extremely good manoeuvreability.
b) Ability to reach over obstructions.
c) Easy access to front, sides, rear and overhead.
d) Large reach for small floor area.
e) Slim design allowing easy integration into restricted workplace layouts.
f) Fast operation due to rotary joints.
g) Ability to traverse complex continuous paths.

Three implementations of the articulated configuration are illustrated in Fig. 2/3.

2.1/4 Polar or spherical configuration

The geometry of the **spherical configuration** combines rotational movement in both horizontal and vertical planes with a single linear (in/out) movement of the arm. It may occasionally be referred to as a *'gun turret' configuration*. Whilst this configuration occupies and sweeps out a relatively large volume, the access of the arm within this total volume is restricted. Particular advantages of this configuration include:

a) Easily controlled/programmed movements.
b) Familiar polar coordinates easily understood.
c) Large payload capacity.
d) Fast operation.
e) Accuracy and repeatability at long reaches.

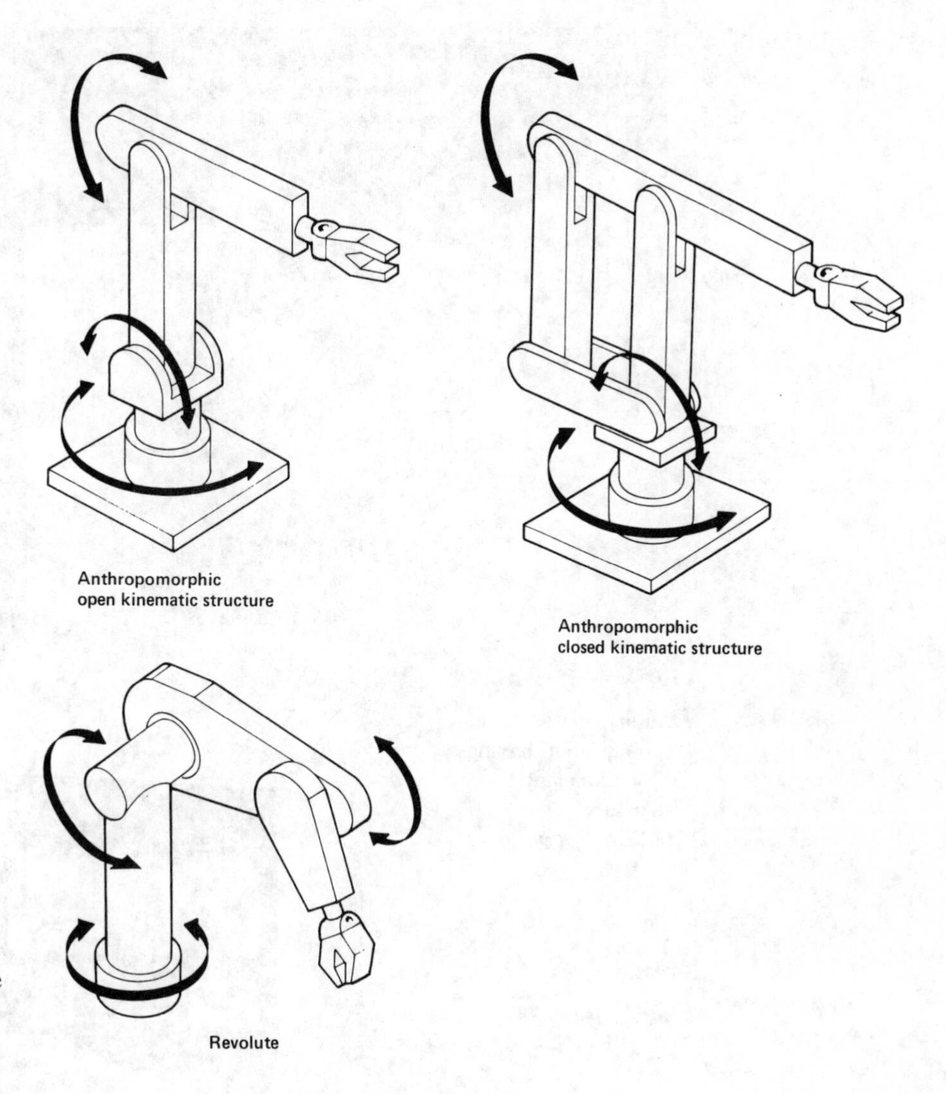

Fig. 2/3 Three implementations of the **articulated** robot configuration

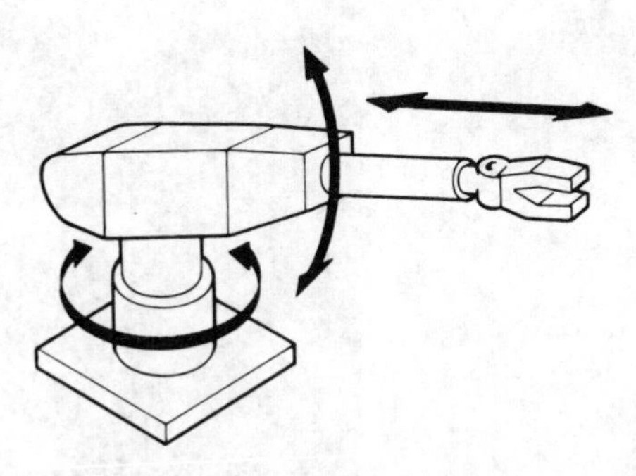

Fig. 2/4 A typical **polar** or **spherical** robot configuration

It is suited to lifting and shifting applications which do not require sophisticated path movements to be traced and is extremely suitable for applications where reaching into horizontal or inclined tunnels may be required. A typical polar configuration is illustrated in Fig. 2/4.

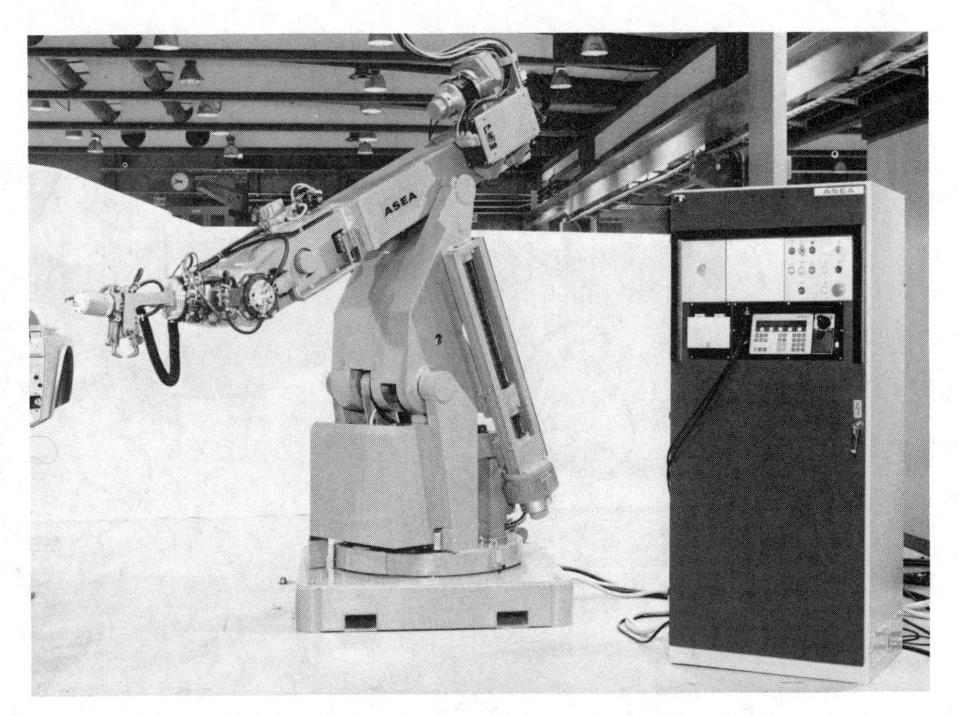

An articulated robot configuration having a closed kinematic structure
[*ASEA Robotics*]

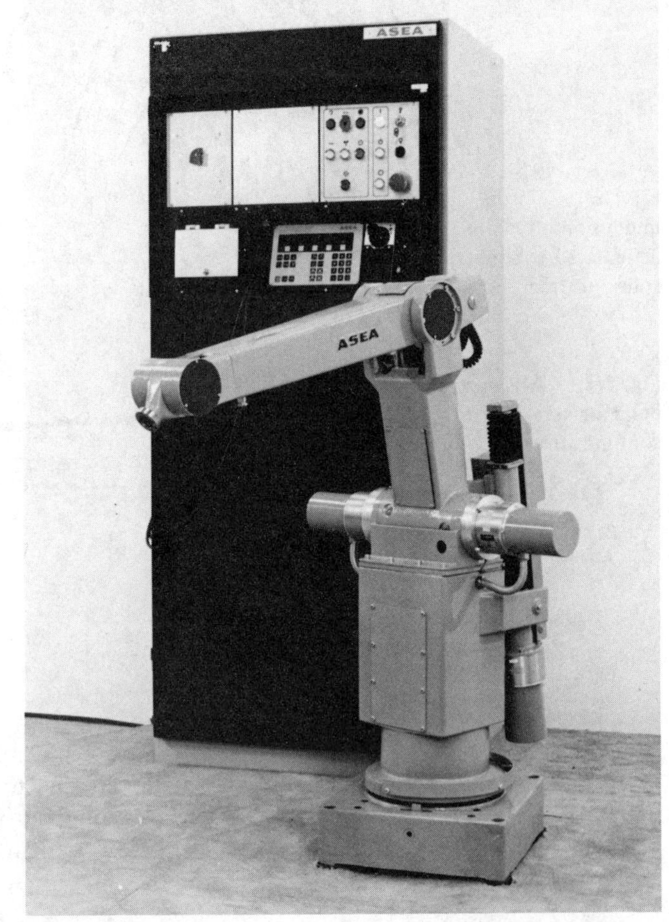

An articulated robot configuration having an open kinematic structure
[*ASEA Robotics*]

2.1/5 SCARA configuration

The term **SCARA** is an acronym for Selective Compliance Assembly Robot Arm. It is a combination of the cylindrical configuration and the revolute configuration operating in the horizontal plane. A three-linked arm with two rotary joints provide movements in the horizontal plane, and vertical movement is provided at the end of the arm. Particular advantages of this configuration include:

a) Extremely good manoeuvreability and access within its programmable area.
b) Fast operation.
c) High accuracy.
d) Relatively high payload capacity due to stiff structure in the vertical direction.

The SCARA configuration was developed primarily for assembly-type operations. Assembly tasks predominantly require movement in the horizontal plane coupled with simple vertical movement for picking, placing and insertion operations. A typical SCARA configuration is illustrated in Fig. 2/5.

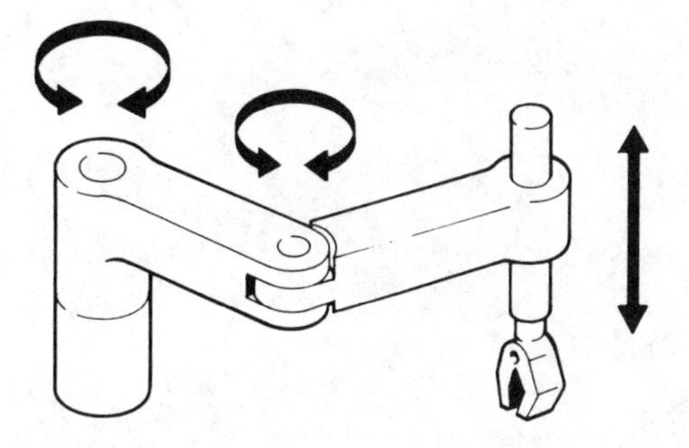

Fig. 2/5 A typical **SCARA** robot configuration

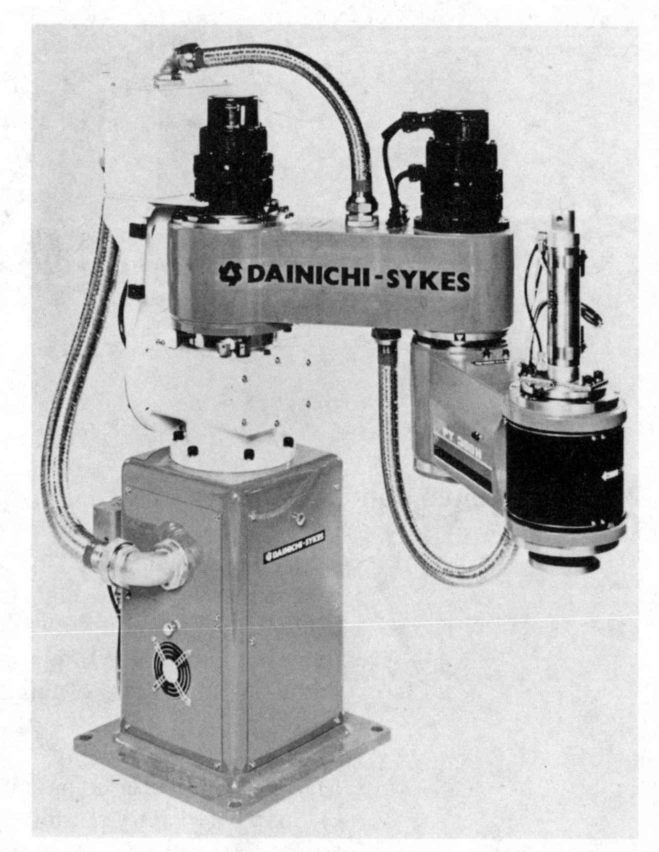

SCARA robot configuration [*DSR Systems*]

2.1/6 Spine configuration

The **spine configuration** appears as a single arm fully enclosed by a flexible protective gaiter. One method of actuation consists of a series of motorised universal joints which can be built up in modular fashion to the required length. A second method of actuation utilises stainless steel ovoids (three-dimensional ovals) strung together on two sets of tensioned steel cables. Linear actuators manoeuvre the arm by altering the tensions in opposing cables. The difficult control coordination in both systems is made possible by sophisticated computer control software. Its relatively low payload capacity is offset by its particular advantages:

a) Extremely good manoeuvreability in restricted spaces.
b) Extremely compact.
c) Ability to reach over, under, round and in and out of obstacles.
d) Relatively small floor space occupied.

This configuration is particularly suited to awkward manoeuvres in restricted spaces (for example, in and out of welded or rivetted body panels) and where close control over movement is required (for example, remote handling of dangerous products). A spine configuration is illustrated in Fig. 2/6.

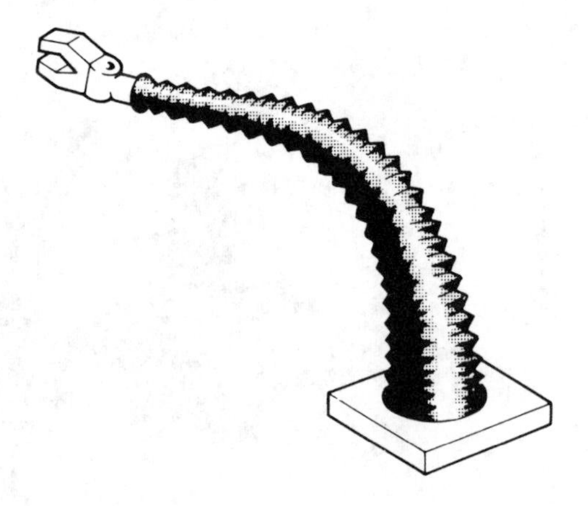

Fig. 2/6 A typical **spine** robot configuration

2.1/7 Pendulum configuration

The **pendulum configuration** combines a single linear motion with two perpendicular rotary movements. The relatively simple design combines fast and efficient operation with high payload capacity. The design is robust, extremely compact and offers great versatility in mounting arrangements.

Particular advantages of this configuration include:

a) High payload capacity.
b) Extremely versatile mounting arrangements.
c) Simple geometry permitting fast operation.

It is suited to heavy tasks where access may be difficult for human operators. One such example is spot welding on the inside of motor car body shells. Mounted overhead the pendulum robot can gain access through the rear window aperture of the body. A pendulum robot configuration is illustrated in Fig. 2/7.

Fig. 2/7 A typical **pendulum** robot configuration

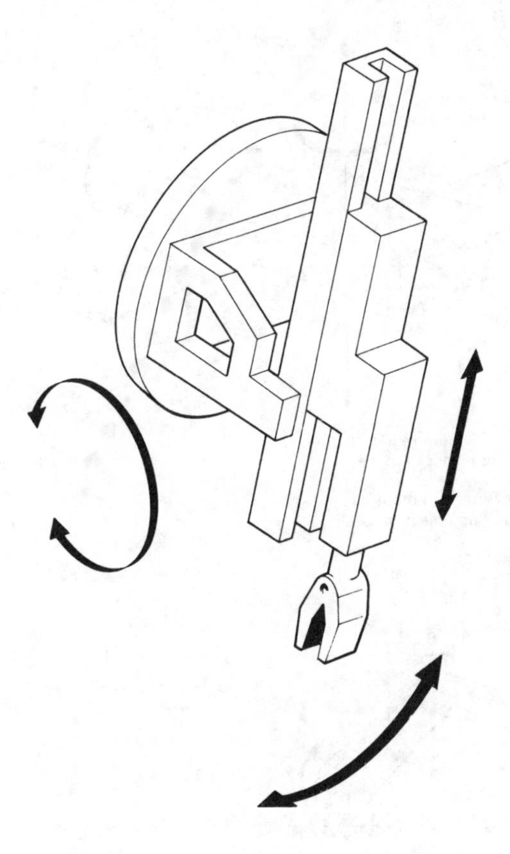

The preceding discussion has intentionally ignored wrist motions. Although many robot types include at least one wrist motion within the basic configuration, wrist motions are dealt with in Section 5/1.

2.1/8 Degrees of freedom and axes of movement

Two terms which are often confused when considering robot configurations are 'degrees of freedom' and 'axes, or degrees of movement'. The distinction is important in robot specification since they relate to different aspects of the robot's capability. The term **degrees of freedom** relates to the locating or positioning of a body in space. Any body in space has *six* degrees of freedom. It can have linear movement along three mutually perpendicular axes and rotational movement about the same three axes. This is illustrated in Fig. 2/8. It follows that the three linear movements (often termed **translational movements**) allow the body to be moved to a desired position in space and the three rotational movements allow the body to be orientated about that position.

In order for a robot to reach all sides of a component, or to take up a position in any attitude, its movements should accommodate the full six degrees of freedom. It is not necessary however for the robot configuration to provide all the degrees of freedom since, in most cases, some (usually three) are provided by the end effector.

Fig. 2/8 The six degrees of freedom of robot movement

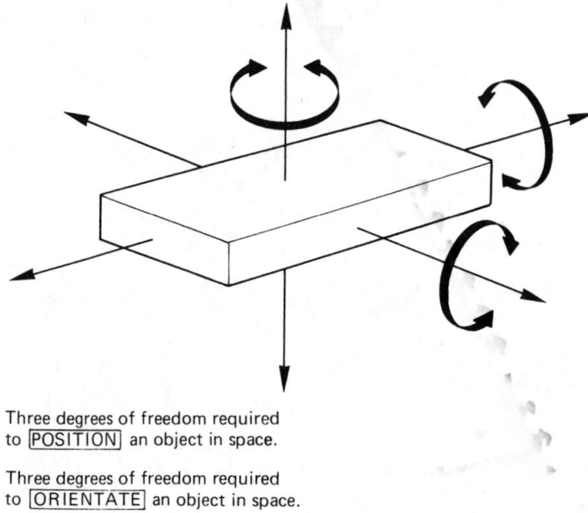

Three degrees of freedom required to POSITION an object in space.

Three degrees of freedom required to ORIENTATE an object in space.

Fig. 2/9 Adding more degrees of movement does not necessarily increase the degree of freedom

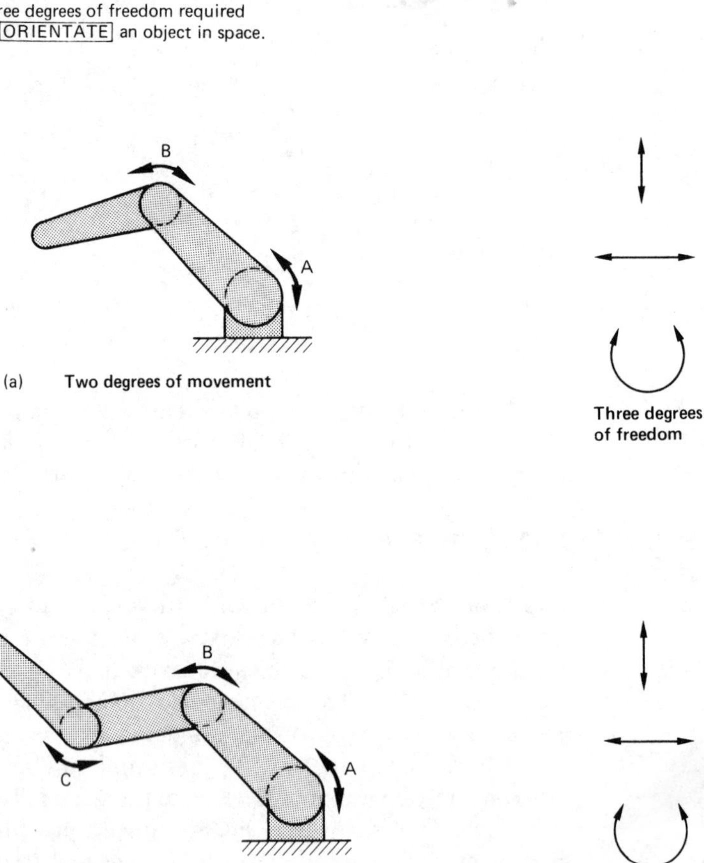

The term **axes of movement** relates to the number of axes in which the robot may move. In one particular robot configuration for example, the axes of movement broadly coincides with the joints of the human body and it is common for them to be termed *waist*, *shoulder*, *elbow* and *wrist movements*.

To demonstrate that the two terms (degrees of freedom and degrees of movement) are not synonymous, consider the simple jointed arm shown in Fig. 2/9*a*. It has two axes of movement represented by the two rotary joints A and B. The joints, when actuated either separately or together, are capable of moving the end of the arm through three degrees of freedom. These are two linear movements and a rotary movement, as indicated by the arrows. Furthermore, adding more joints to the arm, as shown in Fig. 2/9*b*, does not necessarily increase the degrees of freedom. Adding more joints to a robot increases its degrees of movement and may increase its reach and accessibility. It may also extend its working envelope but it can also make programming and control more difficult.

Note that the addition of a gripping motion to an end effector does not constitute either a degree of freedom or a degree of movement.

For the robot configurations described, the following observations can be made. Note that they relate to the basic configuration of the robot and the capabilities of a typical control system. The observations exclude any consideration of an end effector.

Robot Configuration	*Degrees of Movement*	*Degrees of Freedom*
Cartesian	3	3
Cylindrical	3	4
Revolute	3	4
Polar	3	3
SCARA	3	3
Spine	multiple	6
Pendulum	3	3

An end effector typically provides up to three additional degrees of movement (termed **pitch**, **roll** and **yaw**). Thus, all robots, when fitted with a suitable end effector, can achieve the full six degrees of freedom which enable them to fully position and orientate a body in space.

2.1/9 Revolute and prismatic joints

It is evident that arm movement can be accomplished by sliding motion, or rotary motion, or a combination of the two. Reference to the illustrations of the various robot configurations will confirm that arm joints are either linear or rotary. When arm movement is accomplished by a linear sliding motion it is said to be a **prismatic joint**. This is because both the cross-section and the linear movements of the joint approximate to a generalised prism. When arm movement is by rotary motion it is said to be a **revolute joint**, because its motion is either a part or full revolution. This is illustrated in Fig. 2/10.

Fig. 2/10 Revolute and prismatic arm joints

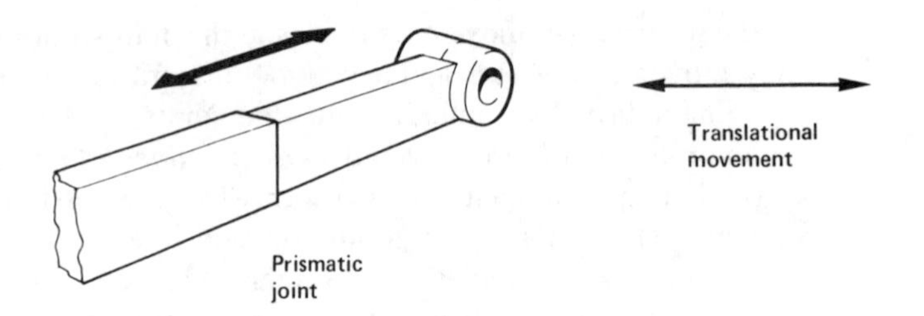

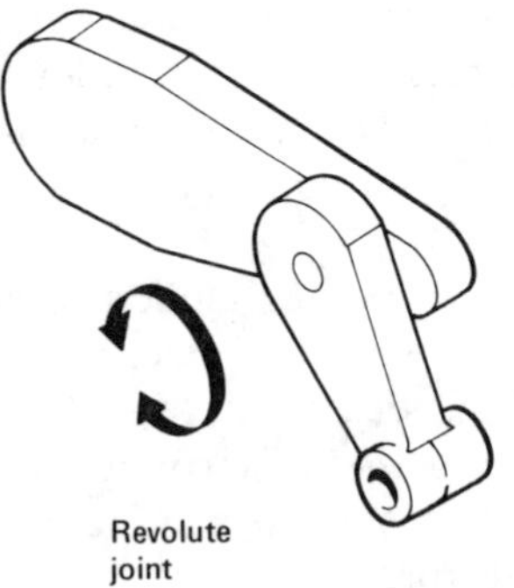

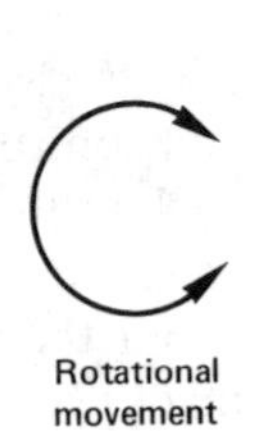

Revolute robot configuration [*Unimation*]

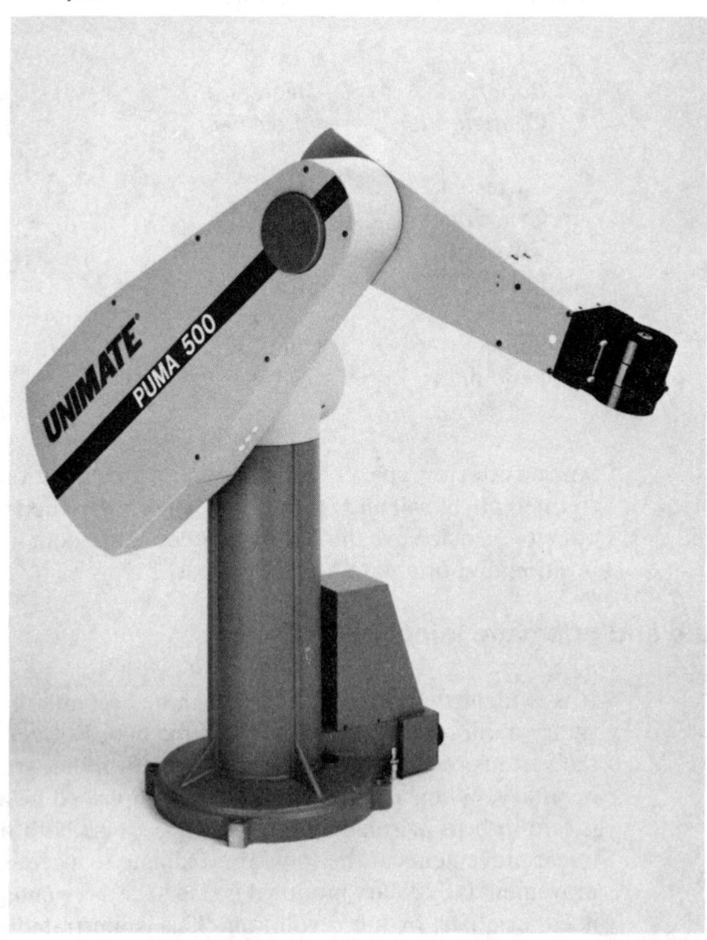

If a prismatic joint is denoted by the letter P and a rotary joint by the letter R, then it is possible to classify a robot by its movement capability. For example, the movement of a cartesian robot configuration could be described as PPP since it consists of three prismatic joints. Some systems, however, designate this movement as TTT, or translational. The common robot configurations can be classified by this method as listed below but the spine configuration cannot be incorporated into this system.

CARTESIAN	PPP	(TTT)
CYLINDRICAL	RPP	(RTT)
REVOLUTE	RRR	
POLAR	RRP	(RRT)
SCARA	RRR	
PENDULUM	RRP	(RRT)

All combinations of arm joint are represented by the common robot configurations. Human motions are achieved by revolute joints only, and, in contrast to most humans, robots (generally) only have one arm.

Physical joint movements are provided by an *actuator*. For example, the linear movement of a prismatic joint can easily be accomplished by a hydraulic or pneumatic cylinder (termed a **linear actuator**) outstroking and retracting. Similarly, the rotary motion of a rotary joint can be achieved by the rotation of an electrical motor (termed a **rotary actuator**), normally with suitable gearing and other transmission elements. There are also ways of achieving rotary motion with a linear actuator and linear motion with a rotary actuator. A rack and pinion arrangement, for example, can accomplish both depending on whether it is the rack or the pinion gear that is driven.

In general, rotary joints can be actuated at greater speeds than prismatic joints. Straight unjointed arms are not capable of reaching over or around objects and obstructions. Physically, rotary joints are easier to seal from the ingress of dirt and other contaminants. Familiar anti-friction ball and roller bearings provide smooth motion to the joints and these can be neatly and efficiently sealed by bearing seals mounted within the bearing housing. It is common to employ angular contact taper roller bearings in revolute joints for a number of reasons, notably that they are designed to carry both radial and axial loads and that they are adjustable, so that the required stiffness and end play can be individually set up on assembly.

Prismatic joints may also use anti-friction slideway elements but, because of their extended length of travel, sealing is more awkard. A common solution is to shroud the slideway in a rubber or fabric **gaiter**. The gaiters are commonly designed on the principle of a concertina to enable them to compress and extend with the movement of the joint. Typical anti-friction joint and slideway elements are shown in Fig. 2/11.

Industrial robots fitted with a number of revolute joints are, by combining their movements, capable of achieving linear degrees of freedom. This is illustrated in Fig. 2/12. A combination of prismatic joints, by contrast, is not capable of providing rotational degrees of freedom, although it can trace circular path movements. A component moved through a circular path traced by linear joints will merely change its position in space, but not its orientation.

Fig. 2/11 Typical anti-friction joint and slideway elements [*courtesy: ESE Ltd.*]

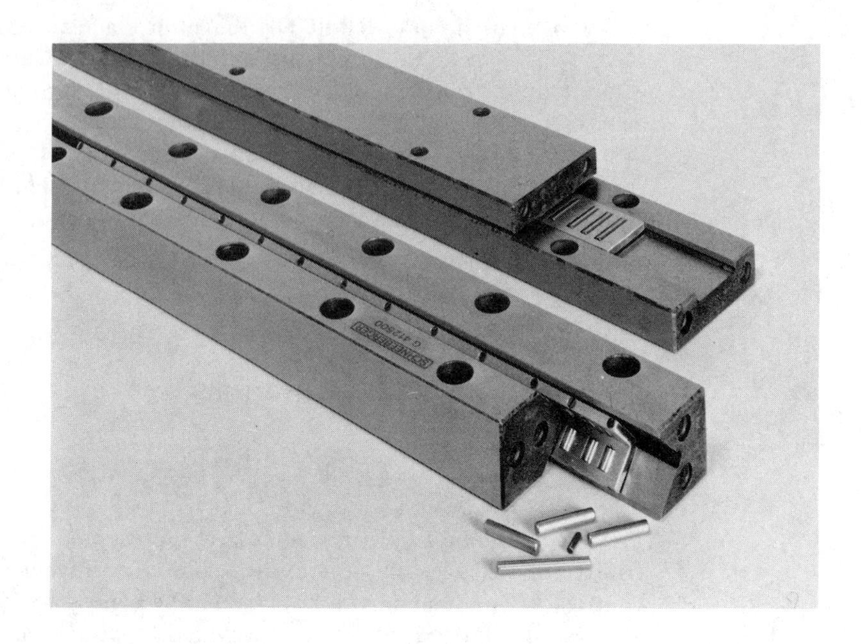

Fig. 2/12 Revolute joints are capable of straight-line motion

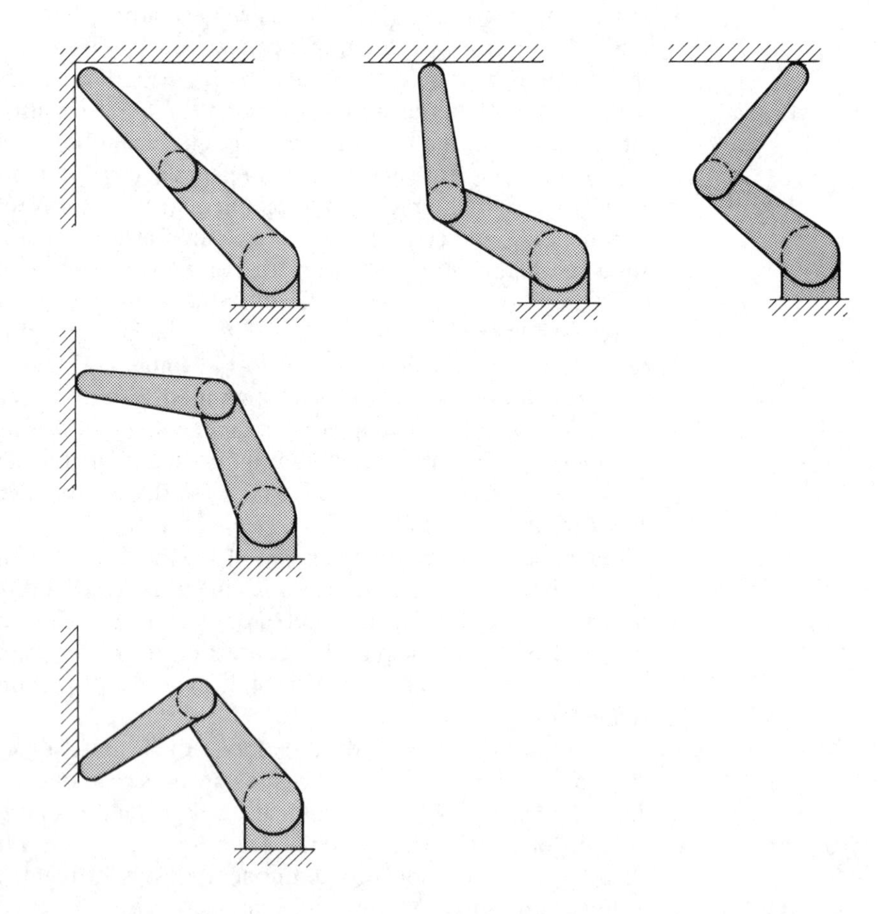

Care must be exercised in the interpretation of robot specifications since they may be influenced by the type of joint employed. For example, remembering that velocity is defined as distance per unit time, then since the distance traversed by a rotating arm varies with its radius, the quoted velocity must relate to a specific arm radius. Velocity also varies with payload and, if movements are small, then quoted velocities may not be attained. Repeatability and accuracy of a revolute joint are a function of the angular displacement of the joint. Thus, quoted linear values may not be appropriate. Since three of the most important aspects of a robot specification are payload capacity, repeatability and accuracy, the above points may be important.

2.2 Working envelopes

2.2/0 Locus and loci

A **locus** is a technical drawing term for the shape of a path traced out by a point constrained to move under certain conditions. Consider a length of cord secured at one end such that it is free to rotate about its fixing. If a pencil is affixed to the free end, and the cord pulled taut, a circle is traced out when the cord is rotated about its pivot. A *circle* is the locus of a point that moves around, and remains equidistant (i.e. at a fixed distance) from, another fixed point. Similarly, an *ellipse* is the locus of a point that moves around two other fixed points such that the sum of the distances from the two fixed points remains constant. These two loci are illustrated in Fig. 2/13*a* and *b*.

Loci is the plural of locus. Loci find application in engineering mechanisms where it is necessary to define the movement paths of various mechanism elements. They are essentially two-dimensional representations of path movements. One mechanical axis drive mechanism that uses the principle of a third locus is the cyclo drive (see Chapter 3, Section 3.4/3.). The locus employed is called a cycloid. The **cycloid** is the locus of a point on the circumference of a cylinder when the cylinder is rotated, without slip, along a flat surface. It is illustrated in Fig. 2/13*c*.

The end point of a robot arm also traces out a path when it is caused to move. It will be capable of traversing a great number of paths (in three dimensions), thus producing many different loci. The arm is constrained to move in paths determined by its physical and geometrical configuration. The sum of all the

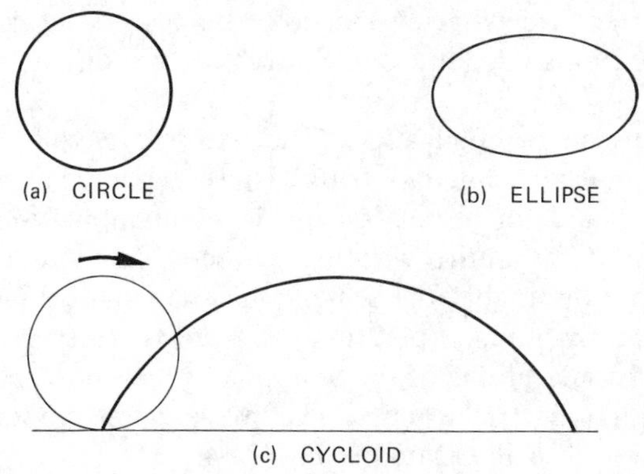

Fig. 2/13 Three common loci: the circle, ellipse and cycloid

possible loci which the robot arm is capable of generating combine to form a three-dimensional working volume called a **working envelope**.

2.2/1 Standard working envelopes

Different robot configurations generate characteristic working envelope shapes. The working envelope is important in selecting a particular robot for a particular task since it dictates:

a) The shape and volume of the workplace that the robot has access to.
b) The volume of the workplace that the robot does not have access to.
c) The extent of reach (in all directions) of the robot.
d) The potential danger zone, surrounding the robot, for safety purposes.
e) The design or redesign of the workplace layout.

Care has to be exercised when interpreting the working envelope of a particular robot configuration for a number of reasons.

a) It should be understood that a quoted working envelope refers to the working volume that can be reached by some point at the end of the basic robot arm only. This point is usually the centre of the end effector mounting plate. It excludes any considerations of end effectors or tools mounted at the end of the robot arm. This is because robot manufacturers cannot predict the task or the shape and size of end effector that the end user will employ. End effectors may thus increase the effective working envelope of the robot. They may also, depending on their design, create additional dead zones.
b) Depending on the geometrical constraints imposed by the physical design of the robot, there may be areas within the overall span of the working envelope that cannot be reached by the end of the robot arm. Such areas are termed **dead zones**.
c) The maximum quoted payload capacity of an industrial robot can only be achieved at certain arm spans. This may, or may not, be the maximum reach at the extent of the working envelope. Maximum payload capacity should also take into account the weight of the end effector.
d) The maximum working envelope achievable by the physical configuration of the robot may be artificially reduced or restricted by the type of control system software and programming method.

The standard working envelope shapes of the common industrial robot configurations introduced in Section 2.1 are discussed below.

1 Cartesian Configuration The working envelope of the cartesian configuration is a *rectangular prism*. There are no dead zones within the working envelope and the robot is capable of manipulating its maximum payload throughout the entire working volume. The simple linear motions of this configuration enable the rectangular-shaped working envelope to be easily visualised by human operators—a factor that is an effective contribution to safety considerations of the workplace when human operators work side by side with industrial robots. The working envelope of the cartesian robot configuration is illustrated in Fig. 2/14.

Fig. 2/14 Working envelope of the Cartesian robot configuration

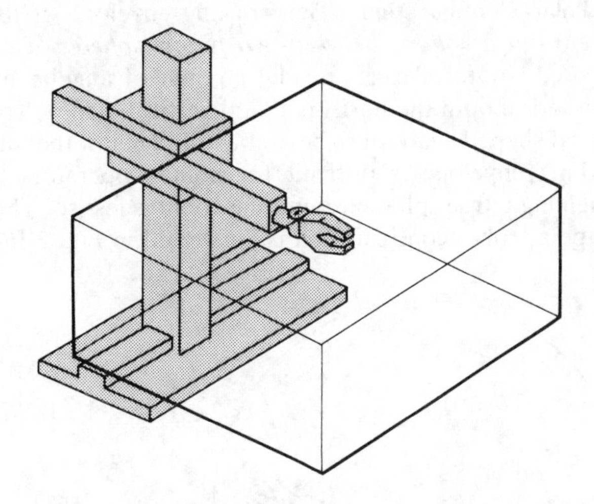

2 Cylindrical Configuration The working envelope of this configuration, as its name suggests, is broadly a *cylinder*. The cylinder is hollow, since there is a limit to how far the arm is able to retract; this creates a cylindrical dead zone around the robot structure. In many cases the working envelope forms only a partial cylinder due to possible constraints on the angular rotation about the base. A truly cylindrical working envelope is only generated if the robot is allowed to rotate, about its base, through 360°. It is not possible for the arm to reach down to the floor or overhead. In some designs, the retracting horizontal arm often extends backwards past the central structure of the robot. It is therefore necessary to recognise (and provide) a clearance volume behind the horizontal arm. As with the cartesian configuration, the working envelope of the cylindrical robot is easily visualised by human operators. The working envelope of the cylindrical robot configuration is illustrated in Fig. 2/15.

Fig. 2/15 Working envelope of the cylindrical robot configuration

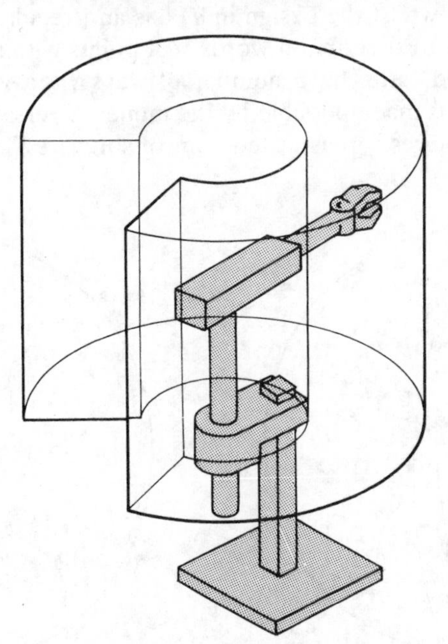

3 Polar Configuration The working envelope of the polar configuration *sweeps out a volume between two partial spheres*. There are physical limits imposed, by the design, on the amount of angular movement that can be achieved in both the horizontal and vertical planes. These restrictions create conical-shaped dead zones both above and below the robot structure. The true working envelope is difficult for human operators to visualise, although imagining a true sphere around the robot is less so. The working envelope of the polar robot configuration is illustrated in Fig. 2/16.

Fig. 2/16 Working envelope of the polar or spherical robot configuration

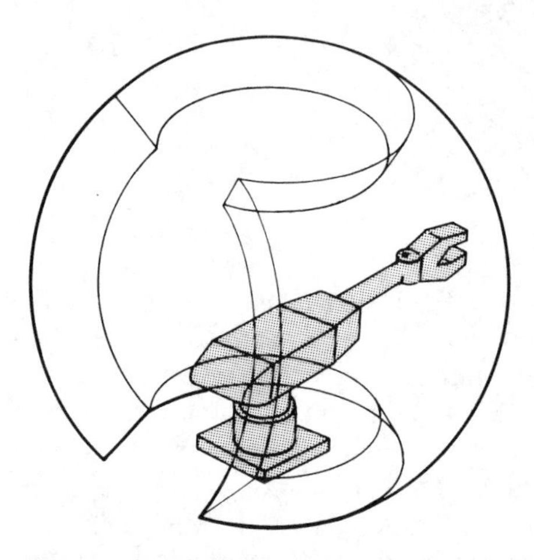

4 Articulated or Revolute Configuration The articulated configuration has a large working envelope relative to the floor space that it occupies. The shape of the working envelope *depends on the physical design*. Two of the most common designs and their associated working envelopes are illustrated in Fig. 2/17. The design in *b*) allows almost a true sphere to be reached (it has a flat circular base), whilst the design in *a*) has an irregular cusp-shaped envelope that is difficult to describe in words. All points within the working envelope can be reached, often in a number of ways via a variety of different arm positions: this is made possible by the immense versatility of the jointed arm design but requires sophisticated control software. This is illustrated in Fig. 2/18.

Fig. 2/17 Working envelopes of articulated robot configurations

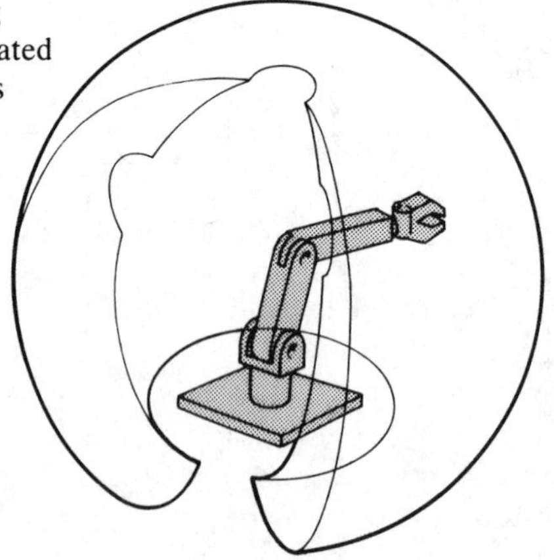

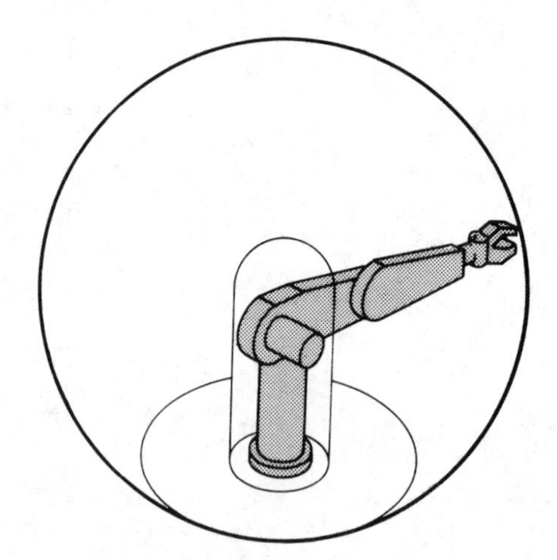

Fig. 2/18 The same position can be reached via a number of arm positions

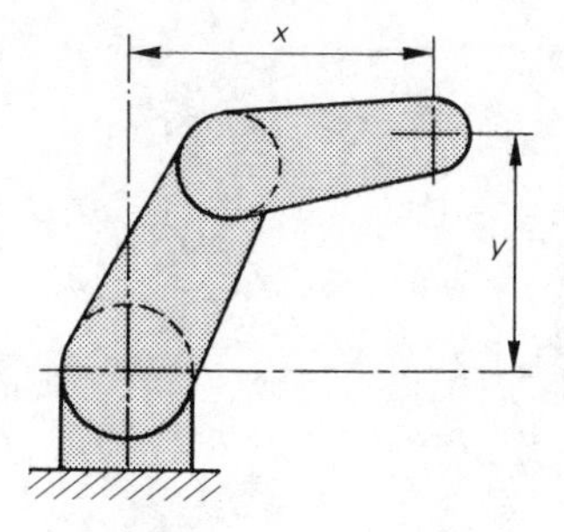

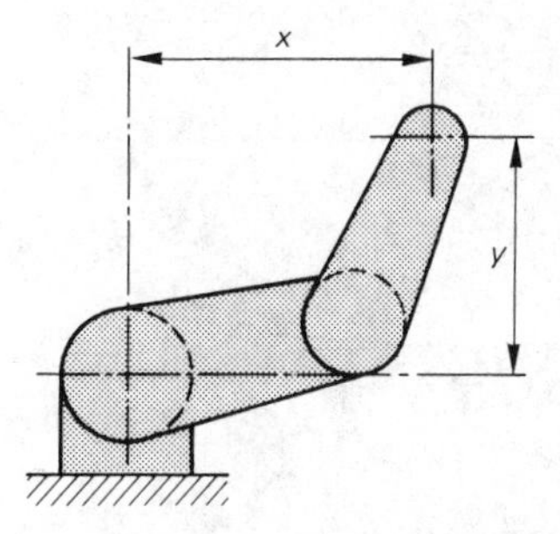

5 SCARA Configuration The SCARA configuration has a working envelope that can be described broadly as a heart- or kidney-shaped prism (in plan view) having a circular hole passing through its middle. This allows for a large area coverage in the horizontal plane but relatively little coverage in the vertical plane. This rather individual working envelope must be viewed in the knowledge that the SCARA configuration was designed primarily for assembly-type tasks in the horizontal plane. It is difficult for human operators to visualise the shape of this working envelope. The working envelope of the SCARA robot configuration is illustrated in Fig. 2/19.

Fig. 2/19 Working envelope of the SCARA robot configuration

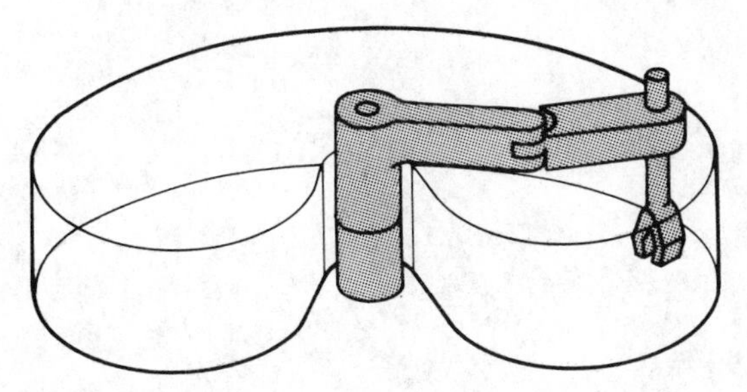

6 Spine Configuration The working envelope of the spine configuration is dependent on its length and the number of articulations provided in its spine. It will *approximate* to that of a true *hemisphere*. The greater the number of articulations, the larger the hemispherical envelope and the closer it approaches the base of the robot. This configuration is characterised by its great flexibility within its working envelope. The working envelope of a typical spine robot configuration is illustrated in Fig. 2/20.

Fig. 2/20 Working envelope of the spine robot configuration

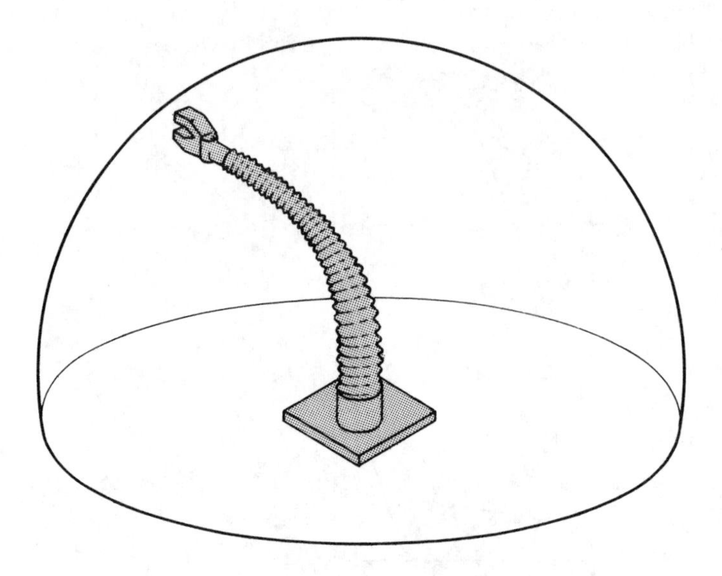

7 Pendulum Configuration The working envelope of the pendulum configuration resembles that of a simple *horseshoe having a segment-shaped cross-section*. The rather limited working envelope is offset by the fact that the pendulum robot can be mounted in almost any position, allowing the working envelope to be finely positioned in relation to its task. The extent of the working envelope may be considerably reduced according to its particular mode of mounting. The configuration is ideally suited to heavy spot welding applications in restricted spaces. The working envelope of the pendulum robot configuration is illustrated in Fig. 2/21.

Fig. 2/21 Working envelope of the pendulum robot configuration

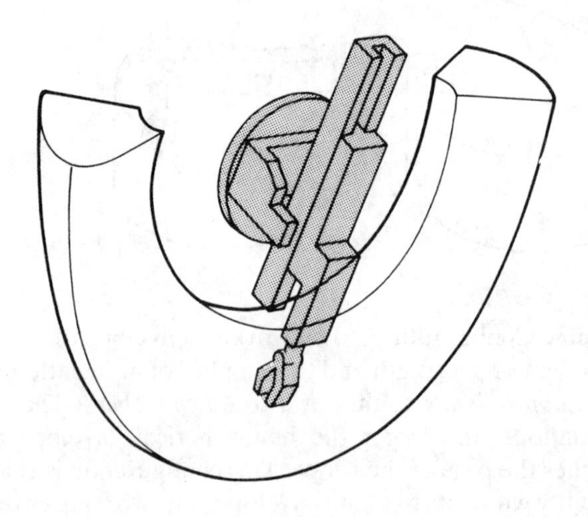

2.2/2 Effect of robot arrangements

The effective working envelope of a robot is usually specified on the assumption that the robot is fixed and floor-mounted in an upright position. Employed in this manner, it is usually termed a **freestanding robot**. This is the way in which most robots are utilised in the industrial environment. It should be noted, however, that this may be imposing unnecessary restrictions on the versatility of the robot. For example fixed, floor-mounted robots:

a) Limit (and fix) the size, shape, disposition and orientation of the working envelope.
b) Tend to be mounted in positions central to the flow of production and can thus be a liability if they fail.
c) May cause congestion in the working environment during maintenance, configuration or programming periods.
d) Discourage changes in plant or workplace layout.

The extent of the working envelope may be re-orientated or re-positioned by considering different **mounting arrangements** for the robot. Robots, unlike human operators, are tolerant of being operated whilst mounted upside down, hanging from the side of an upright structure or mounted on a machine tool. This may offer the following advantages:

a) It may make access to the work task more amenable to the robot.
b) It can release valuable floor space on the shop floor.
c) It can remove a significant obstruction from the workplace.
d) It may improve static, dynamic or load-carrying characteristics.
e) It may reduce arm travel and hence lower cycle times.
f) It may enable a simpler (or cheaper) robot configuration to be specified.

Examples of alternative mounting arrangements in which most robot configurations may be employed include:

a) Machine-mounted.
b) Side-mounted from a rigid structure.
c) Vertically rail-mounted.
d) Horizontally rail-mounted (parallel and transverse).
e) Overhead- or gantry-mounted.

Some of these mounting arrangements are illustrated in Fig. 2/22.

The size of the working envelope may be increased, or the mobility of the robot improved, by providing a particular mounting arrangement with some means of movement. This is termed giving the robot **locomotion**. This may be necessary for the following reasons:

a) It may allow the robot to move into and out of its working position within a congested workplace.
b) It may be necessary to service a dynamic work task, for example tracking a moving conveyor or production line.

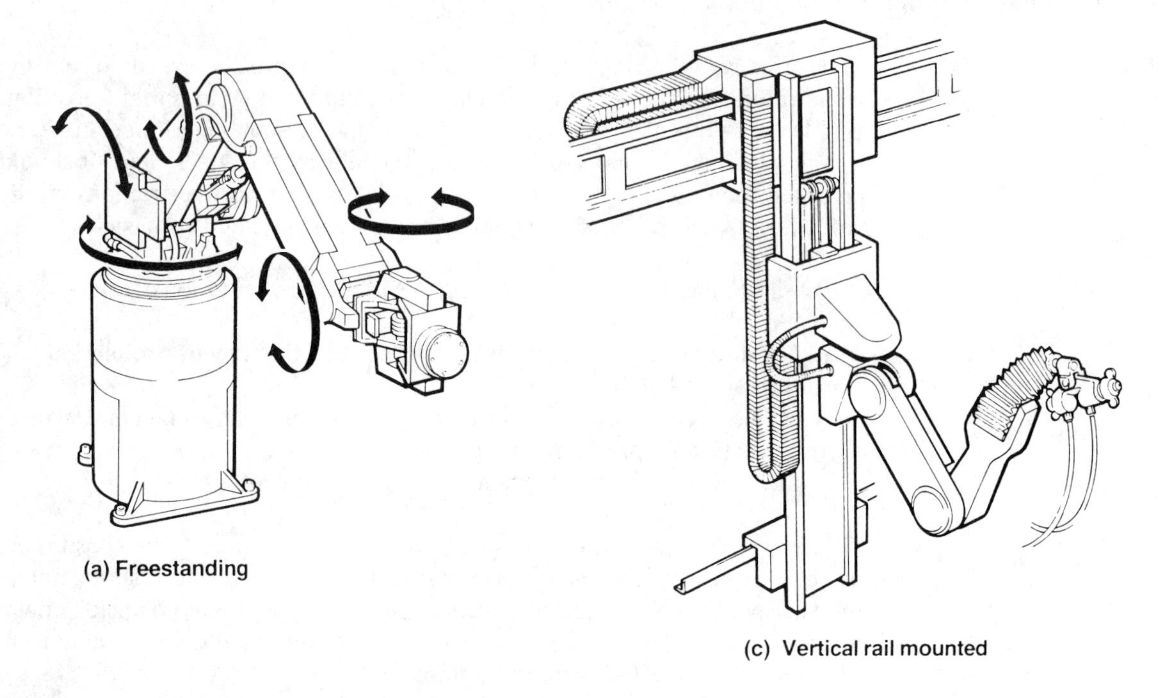

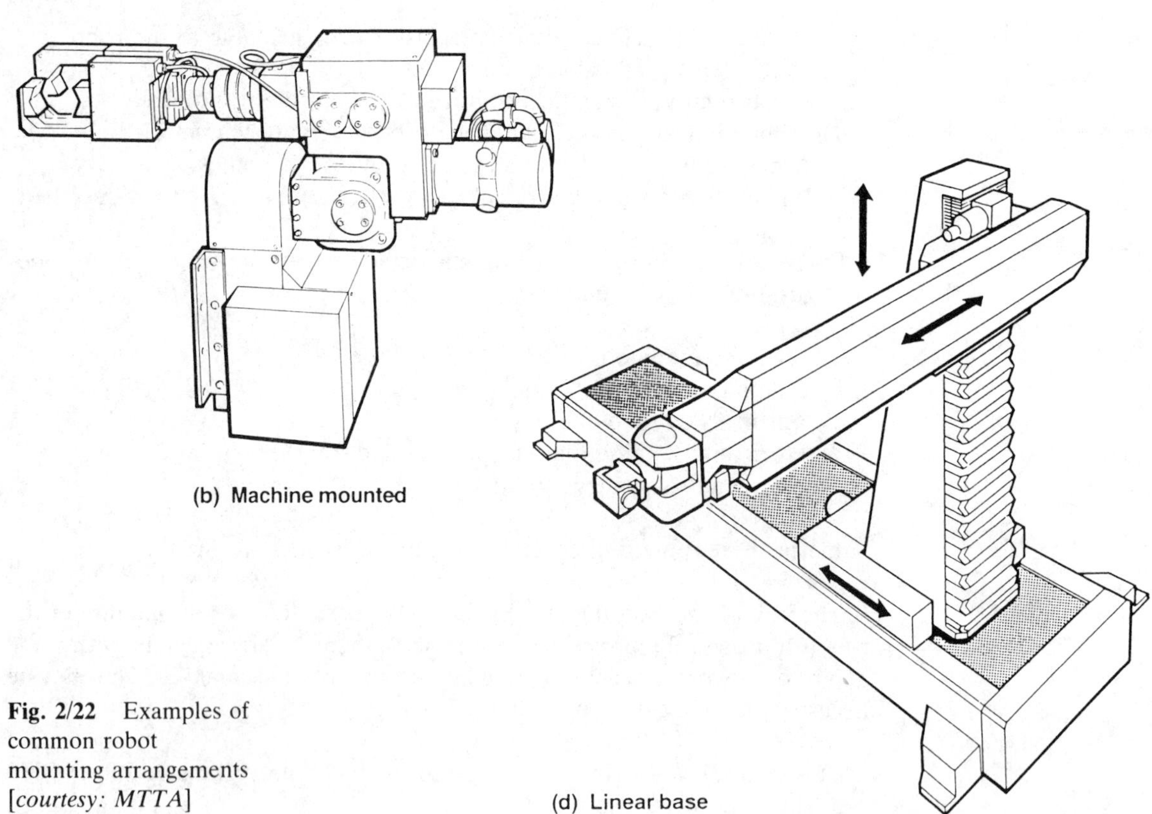

Fig. 2/22 Examples of common robot mounting arrangements [*courtesy: MTTA*]

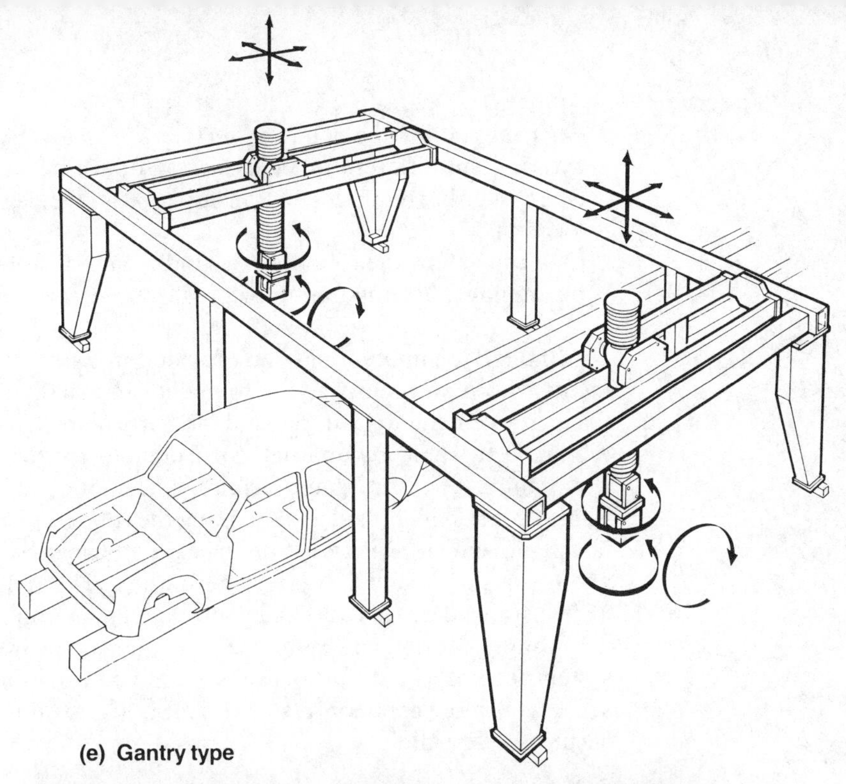

(e) Gantry type

Overhead mounted revolute robot
[*Unimation*]

c) It may allow the robot to service a number of workstations, possibly carrying out different tasks at each station.
d) It may be required to transfer cargoes between different workspace locations.
e) It may allow the robot to be withdrawn to a more amenable position for maintenance, configuration or programming purposes.

Although locomotion may be provided in a number of ways, the requirements within an industrial environment are usually simple. Straightforward linear movement in a horizontal or vertical plane is all that is generally required. It would be unusual, for example, to require industrial robots to climb stairs or traverse rough terrains. Locomotion in the industrial workplace is most commonly provided by running the robot on fixed rails, along a precise line of movement. Robots are often heavy and need both support and guidance and it is a simple and cost-effective solution. The rails may run at floor level or be arranged on an overhead gantry depending on the requirements of the application. Motion and motion control must be provided to maintain overall system accuracy, or the robot can be equipped with sensory feedback, docking sensors or, more commonly, simple shot-bolts, to monitor its travel and ensure accuracy of location.

2.2/3 Load carrying capacity and reach

The load-carrying capacity of an industrial robot, commonly termed the **payload capacity**, depends on a number of factors:

a) The motive power used to drive the actuators.
b) The configuration of the robot.
c) The physical size, stiffness and rigidity of the robot.
d) The maximum reach of the robot.
e) The speed, accuracy and repeatability desired.
f) The weight of the end effector.

Generally, hydraulic actuators are capable of giving the best power-to-weight ratio for a given application. There is however a growing trend towards the use of cleaner, more convenient electrical drive systems.

Some robot configurations are inherently stronger and more rigid than others. This enables higher loads to be transported. In general, the more restricted the movements in a particular robot configuration, the greater the load-carrying capacity. There is often a trade-off between manipulative flexibility and load-carrying capacity.

The disadvantage of load-carrying capacity within a particular design configuration can be offset by making the robot physically larger. Larger robots employing larger, more rigid arm elements will obviously be more capable of carrying higher loads than smaller ones. The trade-off here is likely to be slow speed of operation and reduced mobility and flexibility, together with increased cost.

There is no doubt that for humans it is much easier to carry a heavy weight with the arms close to the body, rather than outstretched. The same principle applies to the robot arm. Large, heavy masses carried at maximum reach can produce undesirable turning effects (called *moments*) to be set up about the central structure. This can cause bending and flexing of physical arm and structural elements, and a tendency for the robot structure to be pulled over with a consequent loss of accuracy and repeatability of robot movement. These conditions are compounded when rapid arm motions are programmed. For these reasons care should be taken to ascertain the true maximum load-carrying capacity of the robot, and the points within the total working envelope at which it can be achieved.

Since most robots are sold without an end effector, the weight of the eventual end effector needs to be considered as part of the payload. The true payload capacity will often be the quoted payload capacity less the mass of the end effector. In some applications, the mass of the end effector may turn out to make quite a significant contribution to the total mass the robot is required to manipulate.

As a rough guide to assessing the load-carrying capacity of an industrial robot, the following guidelines are offered:

Light duty	up to 15 kg
Medium duty	between 15 kg and 40 kg
Heavy duty	above 40 kg

The above limits can only be very approximate since the terms light, medium and heavy duty are relative in their definitions. In terms of payload carrying capacity and size, it is estimated that industrial robots can transport loads that equate to approximately one twentieth their own weight.

2.3 Robot specifications

2.3/0 Measures of specification

There is no standard means of specification for industrial robots. Each manufacturer specifies certain features, including performance data, within the technical literature made available for each particular robot. This means that it is difficult (often impossible) to directly compare robots of even the same configuration. Even commonly applied measures of specification (such as payload capacity, accuracy, reach, etc.) may be interpreted differently by different manufacturers. This in turn may be quite different from the end user's interpretation. There is no easy answer to the problem of comparisons between industrial robots. Robot manufacturers must be closely questioned as to what is actually being specified, before valid comparisons can be made. Specifications may relate to either physical characteristics or performance characteristics.

Physical characteristics are less prone to misinterpretation since they relate in the main to factual information. It is relatively easy, therefore, to make reasonably accurate comparisons between robot types on the basis of physical characteristics. They are, however, the least useful in terms of what the robot can actually do.

Physical Characteristics

Physical specifications of industrial robots include some or all of the following:

a) Robot configuration. *b*) Number of axes of movement. *c*) Type of axis movement. *d*) Floor space required for mounting. *e*) Weight. *f*) Physical dimensions. *g*) Physical details.	MECHANICAL
h) Power drive system. *i*) Power/services requirements.	POWER
j) Programming method. *k*) Type of control system. *l*) External sensors supported. *m*) Program backing store device. *n*) Memory size.	CONTROL

Performance Characteristics

a) Accuracy. *b*) Repeatability. *c*) Resolution. *d*) Velocity range. *e*) Operating cycle time. *f*) Load-carrying capacity.	SPECIFIC
g) Life expectancy. *h*) Reliability. *i*) Maintainability. *j*) Mean Time Between Failure (MTBF). *k*) Mean Time To Repair (MTTR).	NON-SPECIFIC

Performance specifications are usually the means by which robots are judged suitable (or able) for a particular application.

2.3/1 Problems of specification

Performance characteristics may be either specific or non-specific as indicated above. Although considerable differences would be expected in the claimed values of the non-specific characteristics, owing to their rather imprecise nature, the specific characteristics may also prove to be just as imprecise. Considerable care should be exercised in their interpretation especially where they are to be used for comparison purposes. Since robot applications are still in relative infancy, it may be some time before meaningful data is derived concerning reliability, service life and maintainability under service conditions.

Accuracy may be understood as the degree to which an actual position, taken up by the robot, corresponds with the commanded position. A quoted specification, however, could equally well refer to the accuracy of movement of an individual joint or to the accuracy of movement of the end point of the robot arm after a commanded move. Accuracy and accuracy drift can be affected by thermal distortions and, as such, specifications should be qualified, possibly by stating a required warm-up time.

Repeatability is a measure of how the positions of repeated, identical movements correspond. Repeatability is perhaps more important than accuracy in the industrial environment where the robot is 'taught' its movements by being manipulated by human operators. Both repeatability and accuracy specifications will be based on measurements taken after the robot has fully ceased its motion, including all physical vibrations and springback. This may be some considerable time after reaching the commanded time. The extra time may not, however, be available in the working situation and thus the specified values may not properly apply. Furthermore, both accuracy and repeatability may be influenced by the load being transported by the robot and by the arm position. The published values should relate to the worst conditions, i.e. with the arm carrying the maximum specified payload extended at the limit of the working envelope.

Load-carrying capacity (payload) may also reduce as the arm becomes more extended. The quoted values should refer to the load-carrying capacity at full arm extension. The potential user should also probe more deeply into load-carrying capacities specified in terms of low, medium and high. Physical quantities and values will almost certainly be required.

Memory capacity of the control unit may be specified either by the number of program steps, or by the total number of characters that may be stored within the internal memory of the control unit. The difference can be considerable. (Memory capacity is discussed in Chapter 4.)

Questions 2

1. Describe *ten* factors, giving reasons for your choice, that should be considered when selecting an industrial robot.
2. Briefly discuss the main features and the relative advantages and limitations of the cartesian robot configuration.
3. Discuss *two* industrial applications to which the cartesian robot configuration may be successfully applied.
4. Briefly discuss the main features and the relative advantages and limitations of the cylindrical robot configuration.
5. Discuss *two* industrial applications to which the cylindrical robot configuration may be successfully applied.
6. Briefly discuss the main features and the relative advantages and limitations of the articulated robot configuration.
7. Discuss *two* industrial applications to which the articulated robot configuration may be successfully applied.
8. Briefly discuss the main features and the relative advantages and limitations of the polar robot configuration.
9. Discuss *two* industrial applications to which the polar robot configuration may be successfully applied.
10. Briefly discuss the main features and the relative advantages and limitations of the SCARA robot configuration.
11. Discuss *two* industrial applications to which the SCARA robot configuration may be successfully applied.

12 Briefly discuss the main features and the relative advantages and limitations of the spine robot configuration.

13 Discuss *two* industrial applications to which the spine robot configuration may be successfully applied.

14 Explain, with the aid of neat sketches, the difference between the terms 'degrees of freedom' and 'degrees of movement' when applied to industrial robot configurations.

15 Define the terms 'prismatic' and 'revolute' in the context of industrial robot arm joints, and discuss their relative features.

16 Define the term 'working envelope' and explain its importance when considering robot specifications for industrial tasks.

17 State *four* alternative mounting arrangements to that of the usual 'freestanding' robot, explaining the advantages to be gained in each case.

18 Explain what is meant by providing industrial robots with 'locomotion'.

19 Outline the important factors that together influence the payload capacity and reach of industrial robots.

20 Discuss, preferably with reference to manufacturers' literature, the various means of formulating robot specifications and performance measures.

Control Considerations 3

3.1 Open and closed loop control

3.1/0 Introduction to automatic control

Automatic control of robot motion is an essential requirement of any robotic installation. Without it, actions would be limited to simple manoeuvres. But how is this automatic control achieved? Consider first some fundamental factors relating to control systems in general.

For our purposes, a **control system** may be defined as:

> *One or more interconnected devices which work together to automatically maintain or alter the condition of a robot element in a prescribed manner.*

Such a system may be mechanical, electrical, electronic, hydraulic or pneumatic. In practice, many control systems are combinations of these and are termed **hybrid systems**.

In theory, an **input signal** is generated in response to an inputted program command. This produces an **output signal** which activates an actuator, which then moves the robot element. In practice, however, to achieve this satisfactorily can be a complex problem.

One important distinction that must be made in relation to control systems is between *open loop* and *closed loop* operation. Consider the following example.

Suppose we have a furnace heated by an electrical heating element and controlled by a dial graduated in degrees C. We require a furnace temperature of 100° C and set the dial accordingly. This represents an *input command* and generates an input signal (voltage) to the heating element. In turn this input signal produces an output (electric current, and hence heat) which controls the final temperature. The temperature is known as the *controlled quantity* and will eventually stabilise at a *steady state* value. If the dial has been calibrated correctly, this steady state value will be 100° C. On the face of it we have a simple control system for furnace temperature as illustrated in Fig. 3/1.

However, consider what happens if the door of the furnace were left open. The temperature obviously drops, yet the input signal (dial reading) and the

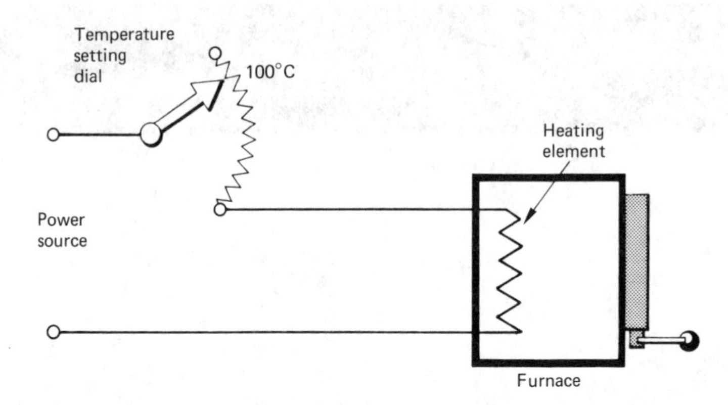

Fig. 3/1 Simple control system for furnace temperature

output signal (current) behave as though they were delivering the requested 100° C. Similarly, if we were to place a red hot casting into the furnace and close the door, the temperature would rise considerably. Again the dial and the current would still behave as though they were delivering 100° C.

In this system even if the temperature in the furnace is unsatisfactory (incorrect), it can in no way alter the input to the furnace control to compensate. Or, put into control terms, *the output quantity has no effect on the input quantity*. In such a case the system is identified as an **open loop control system**. The adoption of open loop control requires very careful consideration since, as illustrated, any change in external conditions may cause the output of the system to fluctuate, or drift, in a manner that cannot be tolerated by the application concerned. It would be intolerable, for example, for a robot control program to indicate a movement of 100 mm and only actually move, say 99 mm. This could happen using open loop control.

3.1/1 Block diagrams

By convention, control systems are represented on paper by **block diagrams**. This allows any system, regardless of power requirement, to be visualised simply and clearly. It is often known as the **black box** approach since a detailed knowledge of the workings of the component parts of the system is not required. It is only necessary to know how the output signal will respond to a given input signal and not what actually happens inside the box.

The block diagram of the open loop temperature control system described above is shown in Fig. 3/2.

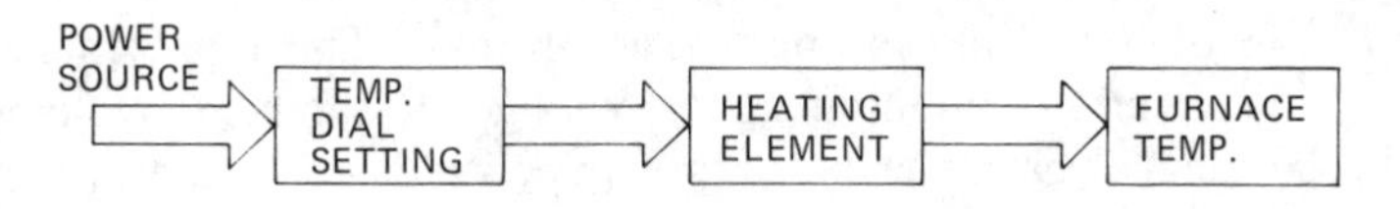

Fig. 3/2 Block diagram of an open loop temperature control system

3.1/2 Feedback

A thermometer (pyrometer) can be added to the system for the purpose of indicating the value of the temperature of the furnace. A human being can then read the thermometer and adjust the temperature dial to increase or decrease the indicated furnace temperature as desired. We have now introduced *feedback* into the control system by specifying and building in a **feedback loop**.

The *output quantity* (temperature) *is now having an effect on the input quantity* (although only manually). This system is now classified as a **closed loop control system**. More precisely, the operator reads the thermometer, compares the reading with the requested value, and compensates accordingly.

This is still manual control. To supply automatic control the thermometer can be replaced with a thermocouple. A thermocouple is a device that produces a voltage proportional to temperature. This voltage is an ideal form of feedback since it can easily be sensed and measured by simple instruments. Since we already have a reference voltage (the original input signal), it should be possible to compare them electrically. If there is a difference between them, then we can assume that the actual temperature is different from the commanded temperature.

The actual control of the heating element thus depends on the *error* or difference between the desired and the actual temperature. The system is said to be **error-actuated** and, since the actual value is subtracted from the desired value (to determine the error), it is said to employ **negative feedback**. That is, we have an automatic closed loop control system employing negative feedback. A block diagram of this system is shown in Fig. 3/3.

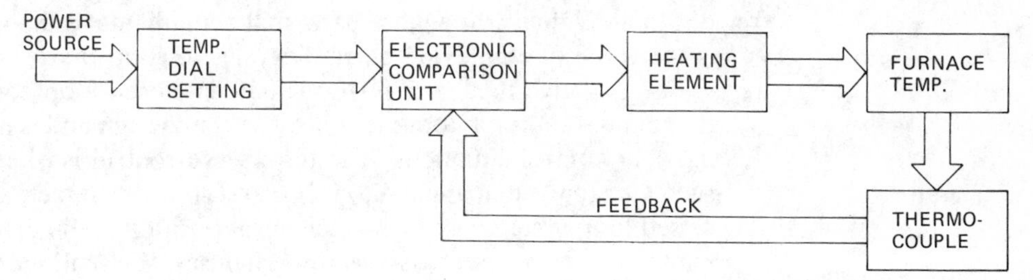

Fig. 3/3 Block diagram of a closed loop temperature control system

3.1/3 Closed loop robotic control

If this principle is now applied to a robot axis movement, a program instruction becomes the command signal, an axis motor becomes the controlled device, and the arm or axis position becomes the controlled quantity.

In physical terms, the command signal itself is unlikely to be large enough to 'drive' an axis motor and would, in practice, have to be magnified by some sort of amplifier. The amount of this magnification is termed **gain** or, more correctly, *loop gain* and becomes very important in control system design. There will, in the case of a closed loop system, also need to be some means of monitoring axis position and some means of comparing input and output quantities, i.e. providing feedback. Fig. 3/4 shows simplified block diagrams of typical open and closed loop control systems as applied to a controlled axis on an industrial robot.

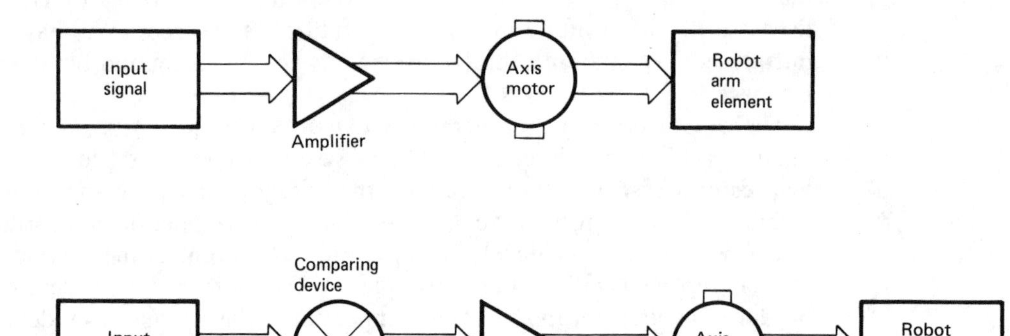

Comparing
device
Input
signal
+
-
Amplifier
Axis
motor
Robot
arm
element
Feedback loop
Position
measuring
device

Fig. 3/4 Block diagrams of typical open loop (*top*) and closed loop (*bottom*) control systems for robot axis drives

Already we have built up a simplified model of a closed loop control system for a single axis of a robot, and yet we do not require any detailed knowledge of the individual component parts that actually make the axis function. Such is the value of employing block diagrams in control system design.

The negative-feedback (error-actuated) closed-loop concept has become the foundation for automatic control system design and is now widely applied in robot control situations. The term **servo-control** is often used to describe such a system when applied to robot axis control. In fact, any fully automatic closed loop system in which mechanical position is the controlled quantity is known, more precisely, as a **servomechanism**. We shall see too, however, that open loop control systems also have their part to play in robot control applications.

Closed loop systems, of necessity, require more component parts, and extra control circuitry, in order for them to perform the feedback function, and are consequently more complex than their open loop counterparts. This inevitably means higher costs for the design and implementation of closed loop systems on industrial robots. We would expect that the performance (quality) of the robot would be correspondingly better to justify this increase in capital investment.

3.2 Fundamental problems of control

There are a number of fundamental problems associated with designing control systems that are fully automatic. The way in which these problems are overcome will determine, to a great extent, the final performance or 'quality' of the robot. This section examines some of the considerations associated with these problems.

3.2/0 Accuracy

It is appreciated in engineering that nothing can be perfectly accurate. This is demonstrated by the fact that most engineering components are produced with reference to both **dimensional tolerances** and/or **geometrical tolerances** (BS4500 and BS308 Part III respectively). These are really statements of how inaccurately we are allowed to work. It is therefore more correct to say that something is accurate to within certain limits, where the limits are specified. A robotic control system is no exception. It shares these same limitations and is only accurate to within certain limits.

True accuracy will only ever be achieved by monitoring and controlling the position of the actual end effector relative to the established datums. Since this is impractical under present-day circumstances, discrete (and often ingenious) **measuring devices** have to be employed, within the design of the robot, on elements that are relatively accessible.

The accuracy of the servo control system therefore depends partly on the accuracy of the measuring device used to monitor the position of the robot axis, partly on how this measuring device is utilised, and partly on any mechanical inaccuracies present within the total system.

For example, it is more accurate to measure the linear motion of a robot arm than to measure the rotation of a leadscrew that may indirectly drive the slide which moves the arm. This is because of losses (such as *backlash* and *torsional wind-up*) that contribute to positional inaccuracies. In both cases the measuring device may be totally accurate by manufacture, but the system accuracy is determined by how they are utilised and other mechanical imperfections.

A typical positioning accuracy specification for an industrial robot could be ±1 mm.

3.2/1 Resolution

The term **resolution** refers to the smallest increment, or dimension, that the control system can recognise and act upon. This is not the same as accuracy. For example, a measuring scale may have 20 divisions engraved upon it in which case its resolution would be 1 in 20. It may not be accurate, however, since the divisions could be unevenly spaced throughout its *range*.

When the range of a measuring device is large, overall resolution is sometimes obtained by utilising a **fine measuring device** (over a small range) backed up by a **coarse measuring device** to permit the full range of slide motion to be measured. This principle is firmly established in the design of the common workshop micrometer. The fine resolution device is the micrometer screw which measures over a range of 25 mm to an accuracy of 0.01 mm. If a greater dimension is to be measured, then a coarse device, the frame, is also employed. The frames of larger micrometers go up in steps of 25 mm. Thus, the resolution of 0.01 mm can be retained over a full range of (large) measurements.

A typical resolution specification for the axis drives of an industrial robot would be 0.001 mm.

3.2/2 Repeatability

As already stated, perfect accuracy is unattainable and so some dimensional tolerance must be applied. The component will be considered 'correct' if its dimensions lie anywhere within this **tolerance band**. If a certain axis position is commanded many times in succession (as with repetitive robotic sequences), there will be a difference, or *scatter*, in the positions actually taken up by the end effector. This scatter is a measure of the **repeatability** of the system. The repeatability of a system is always better than its accuracy.

A typical repeatability specification for an industrial robot would be ±0.2 mm. The repeatability depends on the robot configuration and the size and mass of the end effector.

3.2/3 Instability

Closed loop control tends to make for more accurate performance, since the negative feedback is continually trying to reduce any error to an acceptable level. Under certain conditions, however, this continuous corrective action can lead to instability. **Instability** is the tendency for the system to oscillate about a desired position. A robot arm motion travelling at high speed may have too much inertia to stop immediately it reaches the commanded position, i.e. when the error becomes zero. As a result it will **overshoot** its target and feed back an error signal. This error signal will cause a reversal of the axis actuator to enable the robot arm to try again to reach the commanded position. If the slide then **undershoots** its target, the procedure will be repeated and so on. This condition is often referred to as **hunting**. It should be clear that instability is a function of closed loop control systems only, since it is entirely due to the characteristics of the feedback loop, the loop gain and the response of the system.

3.2/4 Response

The **response** of a control system is the time lag between the application of the input signal and the controlled condition reaching the desired value. The tuning of the response speed for any particular control system is a compromise between a minimal time lag and maintaining the stability of the system. It is a complex problem, since to reach a commanded position from rest a robot axis element first has to accelerate, achieve a **steady state** condition, and then decelerate onto the target with the minimum of overshoot. This has to be maintained under widely varying load conditions, i.e. under rapid or feed motion, with an empty or fully laden end effector, and at various positions within the working envelope of the robot.

3.2/5 Damping

To counter the effects of excessive overshoot or undershoot, and hence help minimise hunting, **damping** may be introduced into the system. A certain amount of damping will already be present within the system due to the effects of friction. Any undesirable oscillations die out more quickly as the amount of damping increases. However, too much damping causes the response of the system to be unnecessarily slow.

The effectiveness of damping is seen in car shock absorbers and swing door closers. More importantly we are probably also aware of the difference in behaviour of such systems should damping be removed!

3.2/6 Control engineering

Control system design is not easy. It relies heavily on the strict applications of complex mathematics and a sound knowledge of the laws of physics. In addition, expertise in mechanics, electronics, pneumatics and hydraulics must also be integrated to produce a working control system.

In fact, since around the 1940s control engineering has matured into an engineering discipline in its own right. The treatment so far has been strictly non-mathematical and has served merely to introduce some of the language, terms and concepts used in modern control system design. It is outside the scope of this book to take the analysis further.

3.3 Types of positional control

One way of classifying industrial robots, other than by their structural configuration, is by the type of positional control applied to their axis movements. This approach defines two broad headings which are sufficient to classify all industrially related robotic equipment.

3.3/0 Positional or point-to-point control

Positional control is employed where the robot axes are required to reach particular fixed coordinate points in the shortest possible time.

To understand the operation, consider the various ways of moving between two points A and B. This two-dimensional example is quoted for clarity. The same arguments can be extended to the third dimension. Fig. 3/5 shows three possible methods.

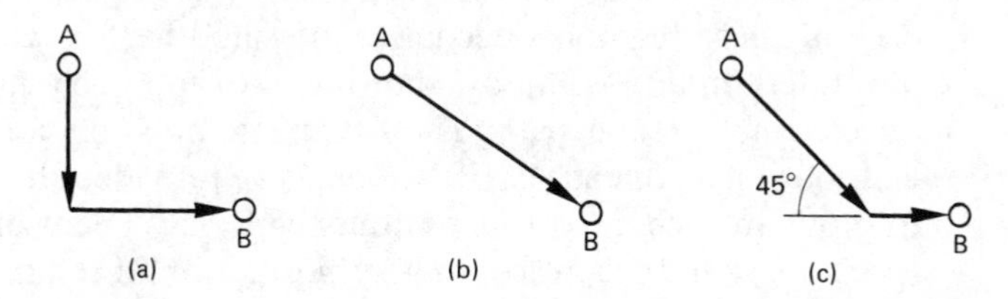

Fig. 3/5 Three possible methods of moving between two points

Method (a) is perhaps the slowest method in that each axis is energised separately. It has the advantage that a very simple control system would suffice since a coordinated movement of the axes is not required.

Method (b) is undoubtedly the quickest route between the points since it is working along the shortest path. This method implies the use of quite sophisticated control systems to coordinate the speeds of each axis in order to maintain a direct line. Control software comprising *linear interpolation* routines is required to trace straight diagonal lines.

Method (c) is a common system, also utilised on point-to-point CNC machine tools. Here, both axes start to move simultaneously at full speed (hence a 45° path), and then to final position in one axis. Control requirements are kept relatively simple without sacrificing too much speed. This path movement, however, could cause collisions with projections that might fall within its programmable area, since its motion can be unpredictable.

Method (c) is the method most commonly used in point-to-point control systems.

3.3/1 Continuous path or contouring control

As stated above, the method by which control systems move from one programmed point to another, when combining axis movements, is called **interpolation**. This is a feature that merges the individual axis commands into a predefined movement path. It operates by calculating intermediate points between given start and end points. When the control system causes the axes to move through these intermediate points, an almost continuous movement is swept. The more intermediate points computed, the smoother the movement obtained. There are three types of interpolation: *linear*, *circular* and *parabolic*. Many robot control systems now provide both linear and circular interpolation. Few controls use parabolic interpolation. For completeness, all three systems will be described.

1 Linear Interpolation This means moving from one point to another in a straight line. With this method of programming, any straight line path can be traced. This includes all diagonal movements. When programming linear moves, the coordinates must be present for the beginning and the end of each line. The end of one line is usually the beginning of the next. The *interpolator* within the control unit calculates intermediate points and ensures that a direct path is traced by controlling and coordinating the speeds of the axis motors. Circular paths or arcs can only be programmed with some difficulty. The circle or arc must be broken down into a number of straight-line moves. The smaller each of these segments becomes, the smoother the contour will be. This is illustrated in Fig. 3/6, by showing the approach in tracing out a two-dimensional circular path. The linear interpolation control requires that the end coordinate of each of the segments be provided. It can be seen that trying to program such curves or contours with only linear interpolation results in specifying hundreds of coordinate positions and requires programs or move sequences of vast length.

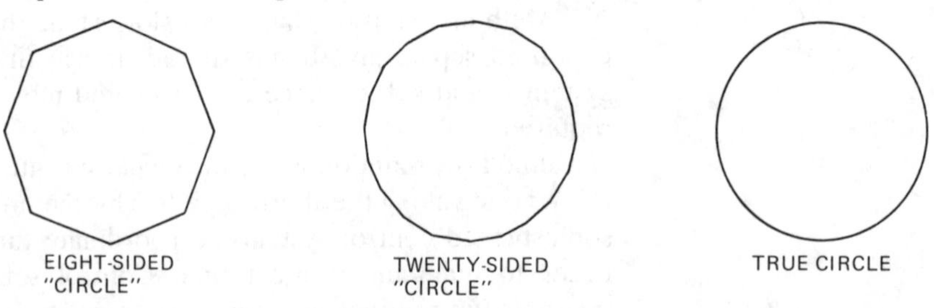

Fig. 3/6 Tracing a two-dimensional circular path using linear interpolation

2 Circular Interpolation The programming of circles and circular arcs has been greatly simplified by the development of circular interpolation. Circular interpolation normally operates from the current programmed position. The end point of the arc and the arc radius must also be specified. The circular interpolation control routine breaks down the arc into small linear moves of high resolution. It is a very versatile function since many shapes can be closely approximated to a series of arcs.

3 Parabolic Interpolation The third method of interpolation is especially suited to free-form contours rather than strictly defined shapes. Parabolic interpolation forms the path between three non-straight line positions in a movement that is either a part or a complete *parabola*. Its advantage lies in the way it can closely approximate curved sections with as much as 50:1 fewer points than with linear interpolation. A parabola is shown in Fig. 3/7.

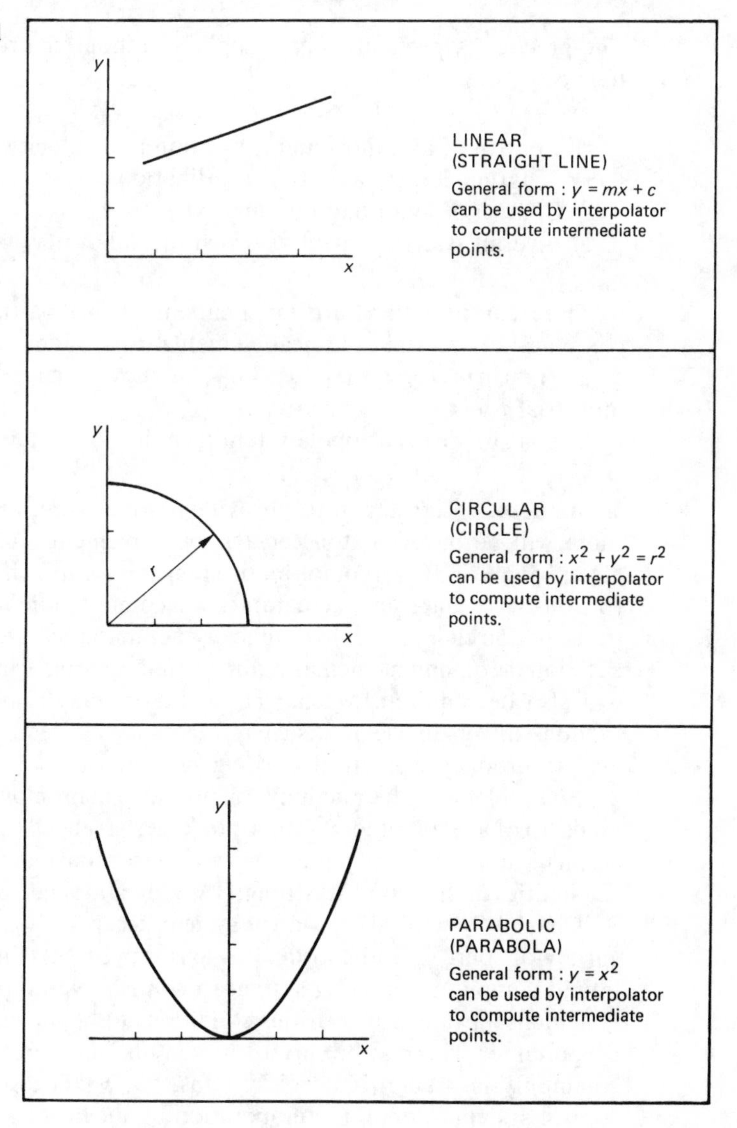

Fig. 3/7 Fundamental basis of linear, circular and parabolic interpolation

Such interpolation methods are ideally suited to the fast and accurate computation abilities of present-day digital computers. Most of the versatility of present-day industrial robots owe their undoubted flexibility to such techniques. Certainly, programming move sequences without them would be a long and arduous task.

All interpolation techniques find their origins in fundamental mathematical analyses of the functions on which they are based. The basis of each interpolation method is illustrated graphically in Fig. 3/7.

3.4 Robot control

3.4/0 Practical aspects of control

The practical aspects of control applied to industrial robots fall broadly into four categories:

a) Control of axis movement, position and velocity.
b) Program control and axis coordination.
c) Control of input and output devices.
d) Overall system control of which the robot may be just one element.

These considerations are independent of any particular robot design or configuration. Certain physical elements can be identified as being common to all industrial robot systems. Fig. 3/8 shows a block diagram of a typical industrial robot system.

A typical industrial robot system includes the following elements:

1 Actuator The actuator(s) provide motive power to the robot. In general, there will be one actuator per axis of movement. Actuator drives may be electrical (AC/DC servomotors or stepper motors), hydraulic or pneumatic (piston drives and rotary actuators), or a combination of these and mechanical transmission elements. If hydraulic or pneumatic actuators are employed, then additional equipment such as pumps, compressors, accumulators, filters, etc. will also be required. Mechanical actuators may also employ intermediate motion-conversion devices such as gears, belts, leadscrews, chains, cables, etc., to produce a practical working system.
2 Manipulator The manipulator provides main movement capability. It is made up of a series of joints and/or linkages arranged to suit a particular robot configuration. The manipulator is also responsible for the load-carrying capabilities of the robot. Its strength, weight and rigidity are important factors.
3 Control System The control system accepts programmed input, from a variety of sources, and coordinates axis movements and velocities. It has an inbuilt computer operating system to control overall operation. The control system also arranges for various conditions to be indicated to the operator or programmer. This may be via indicator lights, audible alarm warnings or, more commonly in modern control systems, via VDU displays. A typical robot control system offers remote operation by means of a *teach pendant*.

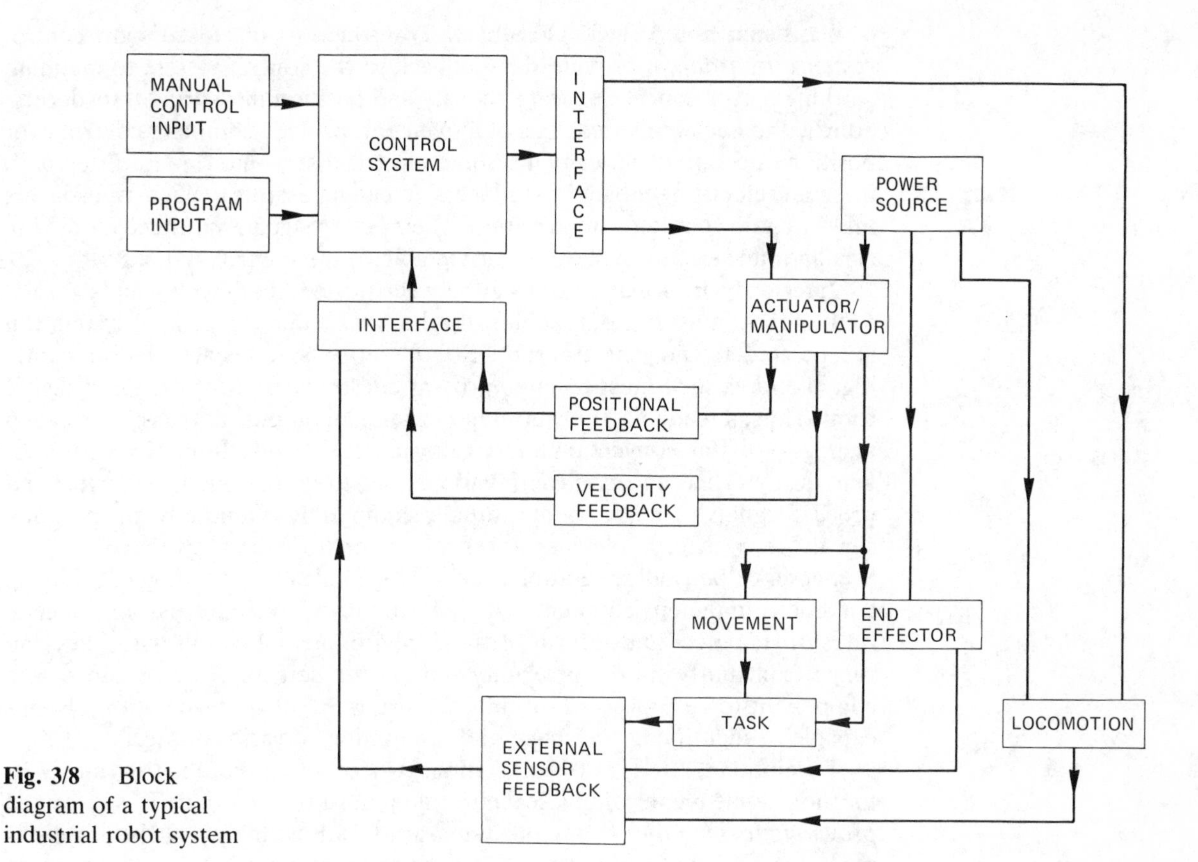

Fig. 3/8 Block diagram of a typical industrial robot system

4 Executive The master control software is often called the *executive* (from the term to 'execute' or carry out). It includes the main operating system of the robot. Second-generation robot control systems are capable of receiving and acting upon external inputs, as well as being able to deliver control signals to initiate or coordinate the response of external devices. These actions, and the priorities assigned to each action, are determined by the executive. It is also capable of responding to sensory information and initiating action accordingly.

5 End Effector The end effector forms the *hand* of the robot. It is not called a hand, since, in many designs, no holding or gripping takes place. Different end effectors allow different tasks to be accomplished by the robot. Since the range of applications to which the robot may be put is diverse, it is common for them to be supplied with some form of universal mounting plate as standard. End effectors of any design can then be fitted, as required, by the end user. They may require their own power supply if they are to be driven in any way. It would be unusual for them to be supplied as standard, and as such they represent an extra cost of installation. (See Chapter 5 for more details about end effectors.)

6 Positional and Velocity Feedback Transducers In closed loop control systems, *transducers* provide the feedback to the control system to maintain working performance. Separate velocity and position measuring transducers, attached to each individual axis of movement, provide complete control over position and rate of movement. Some electrical servomotors are fitted with integral velocity feedback transducers (tachogenerators). Such transducers may be analog or digital in operation. Their output signals may need modifying (via an interface) to make them acceptable to the control system.
7 Interface Information transfer between devices, both within and external to the robot, must be compatible with the signals that the control system can accept, understand, and transmit. Robot control systems are predominantly digital and as such must receive, process and transmit information in digital form. Thus, if feedback transducers, for example, output analog voltages, then analog-to-digital conversion must take place. A collection of electronics, generically called an *interface*, invariably has to be provided to convert and process signals to make them mutually compatible. Generally all incoming signals, from external devices or sensors, pass through an interface.
8 Sensors Second-generation robots are capable of responding to changing conditions in the environment within which they work. External transducers, or *sensors*, detect and inform of these environmental conditions. They can range from fairly simple proximity sensors for detecting the presence of a component to be processed, to fully sophisticated image-recognition systems capable of identifying the shape and orientation of various objects.
9 Locomotion In robot workstations where the robot has to move its location, some means of physical movement must be provided. This is termed *providing locomotion*. Common systems of location employ tracks or rails as means of guidance and support. These may be mounted at floor-level or on a gantry-type arrangement. The executive needs to coordinate the final rest position of the robot with program-sequence execution. If locomotion is not controlled manually, then a docking sensor, positional feedback or shot-bolt system is required to confirm the target rest position of the robot.

The interconnection of these elements is shown in the block diagram of Fig. 3/8. The following discussion introduces those factors that influence control of axis movement, velocity and position in industrial robots. Program control and control of input and output devices are discussed in Chapter 4.

3.4/1 Control of axis movement and velocity

To accurately control positional movement, both positional and velocity information is required. Control of both axis movement and velocity is determined by the type of control system employed. Consideration must be given to both open and closed loop control systems.

By definition, **open loop control systems** do not employ feedback. This implies that neither the movement nor the velocity of the axis motion is being measured. To accomplish accurate control of both movement and velocity in open loop control systems, a special motor known as a **stepper motor** is employed. The stepper motor is unique in that it does not need the application of a varying (analog) voltage, as do conventional AC or DC electrical motors. The stepper motor is a digital device.

The principle of the stepper motor is that, upon receipt of a digital signal (a *pulse*), the motor spindle rotates through a specified angle (the *step*). The step size is determined by the design of the motor but is typically between 1.8 and 7.5 degrees. Thus, if two digital pulses are applied, then the spindle rotor rotates by two steps, or by between 3.6 and 15 degrees, depending on the motor design. By counting (electronically) the number of pulses sent to the stepper motor, a known angle of rotation can be predicted. In practice, a number of electrical coils form the poles of the motor. If two coils are energised simultaneously, the combination of the magnetic field produced detents the rotor in a fixed position. If alternate coils are energised and de-energised in the correct sequence, the rotor takes up corresponding positions, thus causing it to rotate. The principle of operation of the stepper motor is illustrated in Fig. 3/9 which shows a motor comprising four poles.

Fig. 3/9 Principle of operation of the stepper motor

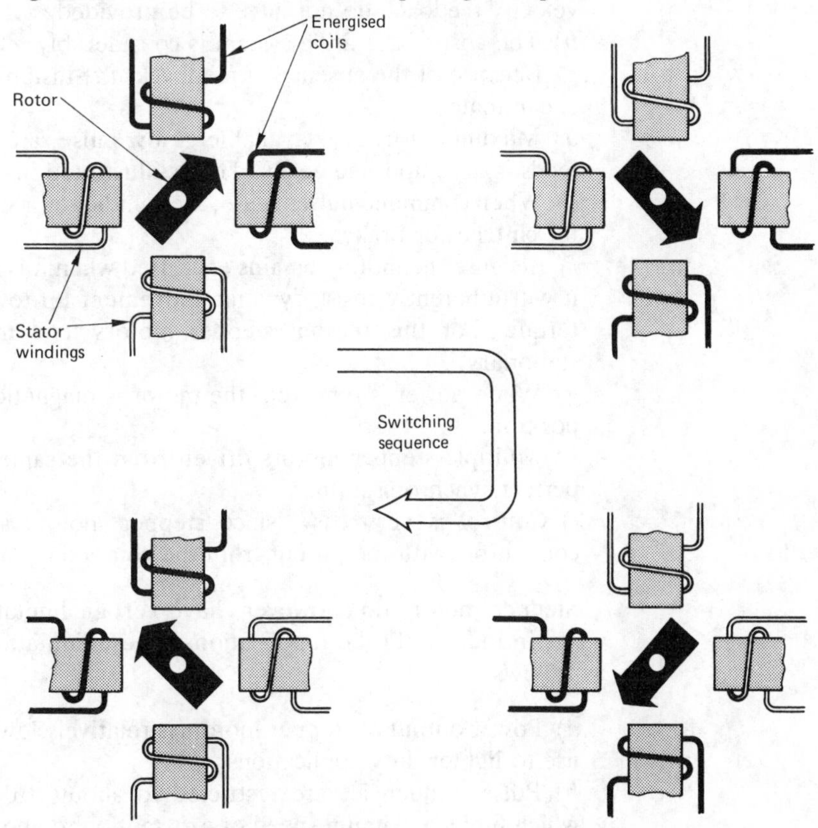

Linear axis movement, if achieved via an axis leadscrew, can be calculated by knowing the lead of the axis leadscrew, the step angle, and the number of pulses. It can be calculated using the simple formula:

$$\text{Linear movement} = \frac{\text{Number of pulses}}{\text{Number of pulses per rev}} \times \text{Lead of leadscrew}$$

Rotational axis movement, if achieved by gear drives for example, can be calculated by knowing the gear ratio and the length of the arm element. There is no need for **positional feedback**.

Velocity of the axis movement is determined by how quickly the pulses are sent to the stepper motor (the **pulse frequency**). If the pulses are sent very rapidly then the axis feedrate will be high; if the pulses are sent very slowly then the axis feedrate will be low. The speed at which the pulses are transmitted can be accurately governed by the computerised robot control system. There is thus no need for *velocity feedback*.

In practice, digital switching circuitry and some means of power amplification are required to drive the stepper motor. This collection of electronics is known as a **translator**.

Advantages of employing stepper motor drives may be summarised as follows:

a) The total drive system is considerably simplified since positional and velocity feedback do not have to be provided.
b) The cost of the drive system is considerably reduced.
c) Because of the absence of feedback, the instability problem of hunting is eliminated.
d) Maximum torque is available at low pulse rates so that acceleration of loads is accomplished easily. This is illustrated in Fig. 3/10.
e) When command pulses cease, the spindle rotor stops and there is no need for clutches or brakes.
f) Because the motor remains energised when it is in a stationary position, it will inherently resist dynamic movement up to the limit of its holding torque. For this reason, stepper motors tend to run hot, even when stationary.
g) When power is removed, the motor is magnetically detented in its last position.
h) Multiple stepper motors driven from the same pulse source maintain perfect synchronisation.
i) Control is very easy, since stepper motors are digital devices and compatible with the output from the computerised robot control unit.

Stepper motors do, however, have certain limitations that restrict their usage in industrial robot applications. These limitations may be summarised as follows:

a) Power output of stepper motors is relatively low, and this restricts their use to lighter duty applications.
b) Pulse frequencies are restricted to about 10 000 pulses/sec (10 kHz) which limits maximum speed of axis motion to about 1/5 of that attainable by closed loop systems.
c) If the axis movement is stalled, the pulses continue to 'count' and loss of position will occur. Furthermore, the control system will not be aware of this loss of position.
d) Ramping-up and ramping-down of pulse rates is required when starting, stopping or reversing to prevent loss of pulse counts.

Closed loop control systems, by definition, require both positional and velocity feedback. Since the control system is not required to count pulses, there is a greater choice of axis drives. AC or DC **servomotors**, or **pneumatic and hydraulic drive systems** can be employed.

Ideal closed loop axis drive servomotors should exhibit the following characteristics:

a) Reversal of direction of movement when required.
b) Torque or force output proportional to speed.
c) High initial starting torque or force.
d) Fast and accurate starts, stops and movement reversals.
e) Mechanical compatibility with the robot design and configuration, and electrical compatibility with the controller.

Mechanical, electrical, hydraulic and pneumatic drive systems, and their associated control mechanisms, are discussed in the following sections. The descriptions are made in comparative isolation for convenience, and it should be realised that many practical industrial robots often combine different systems. It must also be emphasised that mechanical, electrical, hydraulic and pneumatic systems are complete disciplines in their own right. The intention is to inform, in a general manner, of the ways in which typical systems function and of their relative merits and limitations.

3.4/2 Electrical axis drive systems

Electrical axis drive systems may be accomplished by digital- or analog-type electrical motors. The choice will be determined by a number of factors such as the size and quality of the robot, the payload capacity, and the type of control system. Digital stepper motors were discussed in Section 3.4/1. The following discussion relates to the analog (DC and AC)-type motors commonly used on larger industrial robots employing closed loop control systems.

DC servomotors offer high power output relative to their physical size. They combine high torque capability, high acceleration and low inertia for optimum system response. They operate on the principle that a conductor, carrying an electrical current within a magnetic field, is subjected to a force. DC series-wound motors are often used since they can exhibit high starting torque. DC motor control also uses comparatively simple and reliable control electronics, readily available at a relatively low cost. Although direct current is often most conveniently supplied via batteries, large DC motors use AC current which has been converted (rectified) to produce a DC equivalent. This is most convenient, since AC is readily available and a continuous steady supply is guaranteed. Speeds in excess of 3000 rev/min and as low as a few rev/min can easily be achieved without stalling. In principle, the higher the input voltage applied to the DC motor, the faster it will rotate, and the lower the voltage, the slower the motor will rotate. This arrangement (known as *linear control*) is not efficient, especially at low motor speeds.

An alternative approach is to control the time that the voltage is applied to the motor rather than the magnitude. A constant voltage is applied to the motor causing it to rotate. When it is sensed that the motor is rotating too quickly, the voltage is switched off. When the motor speed begins to fall, the voltage is re-applied. Thus, pulses or bursts of input signal are continuously applied to the motor in such a way as to control its speed. The technique is known as **pulse width modulation**. For a high motor speed, the voltage will be switched on for a longer period of time than it will be switched off.

Conversely, for a low motor speed, the voltage will be switched off for a longer period than it will be switched on. When the voltage is switched on, it is known as a **mark time** and when the voltage is switched off it is known as the **space time**. The speed is controlled by these relative time spaces and is known as the **mark-space ratio**. Reversing the voltage polarity causes the motor to rotate in the reverse direction.

Certain DC servomotors can require quite regular maintenance, notably to brushes and commutators, and they are more likely to cause radio interference and be more of a hazard in volatile or explosive environments. DC servomotors for robotic applications normally come in a complete, compact, cased unit comprising the motor, encoder/tachogenerator and fail-safe braking device.

Some brushless design DC motors, which exhibit high-torque/low-speed characteristics, can be used as direct drives. This eliminates the need for, and the problems associated with, mechanical speed reduction elements. Such motors have starting torques some three or four times that of the continuously rated torque capability because they employ rare earth magnets within their construction.

AC servomotors have not been widely applied in robotic applications, because of problems of speed control. In principle, the higher the frequency of the alternating current applied to the motor, the faster it will rotate. Providing varying frequency supplies to a number of axis drives simultaneously has been, until recently, largely impractical.

The fact that both types of motor are electromagnetic is used to provide regenerative braking to cut down deceleration times and minimise axis overrun. Both types of servomotor exhibit an increase in speed with an increase in input signal. However, for any given increase in input signal (either voltage or voltage frequency), the speed of the motor cannot be accurately predicted. Extra arrangements have to be made to measure the speed of the motor and to compare this actual speed with the commanded speed. In control terms, **velocity feedback** must be provided; it is provided by building in, normally within the servomotor case, a device called a tachogenerator. A **tachogenerator** is a device that gives out a voltage proportional to its speed, rather like a simple dynamo. This voltage is used as the feedback to monitor motor speed and hence axis speed or feed.

The selection of an electrical motor for a given application must reconcile the three quantities *power*, *speed* and *torque*, which are related. **Power** is a measure of the capacity of the motor to perform useful work. Larger motors than are theoretically possible often need to be specified depending on the efficiency of the motor. Not all the input energy is converted into useful work. Some of the input energy is wasted by being converted into heat and noise energy. **Torque** is a measure of the strength or turning capacity of the motor. The torque required in a robot application will vary as the arm moves and as the load carried by the arm changes. The weight (force exerted by gravity) of the arm, the weight of the end effector and the weight of the cargo also tend to produce a torque. This may act in opposition to the required direction of motion and needs to be overcome before movement can take place. **Speed** is the link between torque and power. This can be illustrated by considering a car going up a hill. In a low gear, there is low speed but high torque. The car can climb hills more easily in low gears. In a high gear, there is high speed but low torque. But the power of the engine remains essentially constant. With electrical motors, however, a high load that causes the motor to run more slowly may cause it to draw more current and thus affect the power used by the motor. Different motors behave in different ways. The behaviour of a particular motor must therefore be ascertained to determine its suitability for the application concerned. This may be done by reference to a *torque/speed characteristic graph*. Typical graphs are shown in Fig. 3/10.

In the SI system of units, power is measured in watts. In converting from the imperial unit of power, one horsepower is equal to 746 watts.

Fig. 3/10 Typical torque/speed characteristic graphs for electric motors

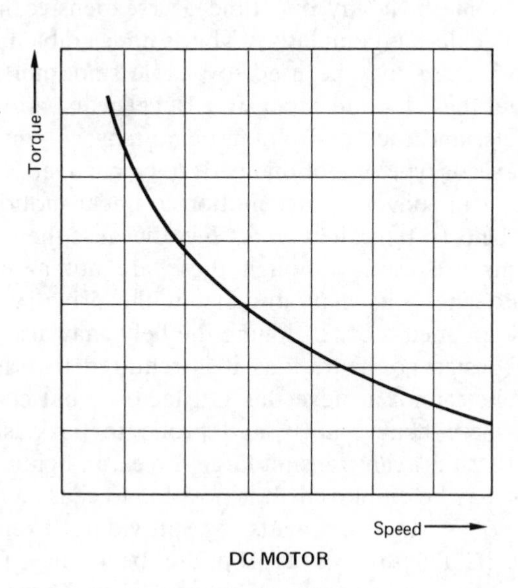

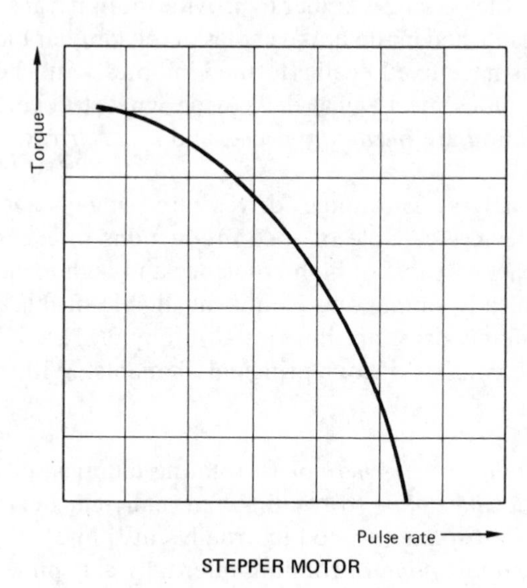

3.4/3 Mechanical axis drive systems

One disadvantage of analog-type electrical motors for robotic applications is that they tend to run too fast. They need to be geared down to reduce their speed and increase their torque. For this reason, they are most commonly employed in conjunction with some mechanical means of power transmission.

Conventional **gearboxes** are not popular since they are heavy, bulky, noisy, prone to inaccuracies through excessive backlash, and may require gear changing mechanisms. Unless single-ratio gearboxes are used (i.e. with only two gears in mesh at any one time), or expensive high-quality gears are employed, backlash is cumulative. This is undesirable in robotic applications. Helical tooth gears may be used to enable smoother transmission and less backlash, but these too are expensive. Bevel gears, crown wheels, worms and worm wheels, and rack and pinion gears may be employed in applications where the axis or type of motion needs to be changed. For example, rack and pinion gears can convert rotary motion to linear motion or vice versa.

Toothed **belts** (often called *timing belts* because they do not slip) and **pulley** arrangements are used, although these are not as effective in high load applications. The belts may stretch but the drive system does tend to be compact. Some means of tensioning the belts may also have to be provided.

In some design configurations it is required to transmit motion around corners. Belts cannot achieve this satisfactorily but chains or beaded cables can. **Chains** are heavy, noisy and prone to backlash but they have the advantage that they can transmit large forces. In lighter applications, flexible steel **cables** may be employed. It is possible to cause a flexible cable to effect movement of various elements by providing fixed beads at intervals throughout its length. An appropriate bead may then be conveniently anchored to the moving member to provide motion transmission. The principle is firmly established in the brake cables of the familiar bicycle. By their design, brake cables have fixed beads (termed 'nipples') attached only to their ends.

Two ingenious mechanical devices designed to overcome the problem of speed reduction are *harmonic drives* and *cyclo-drives*.

Harmonic drives Harmonic drives are single-stage mechanical speed-reduction devices capable of speed reductions in excess of 320:1. They are light, compact, capable of high efficiency and high torque transmission. They are mechanically simple and exhibit negligible backlash. The elements of a typical harmonic drive are illustrated in Fig. 3/11.

The drive consists of four principal elements. With reference to Fig. 3/11 they are

1. The *input wave generator* (a rotating elliptical bearing).
2. The *flexible spline* (a rotating externally-cut flexible spline).
3. The *fixed spline* (a fixed internally-cut spline).
4. The *output spline* (a rotating internally-cut spline element).

The electrical axis motor is keyed to, and drives, the input wave generator. The wave generator rotates inside the flexible spline which is an elastic steel ring with a large number of teeth cut externally on its circumference. The assembly is so constructed that, if the flexible spline is held stationary, the wave

Fig. 3/11 Elements of a harmonic drive speed reducer [*courtesy: Harmonic Drives Ltd.*]

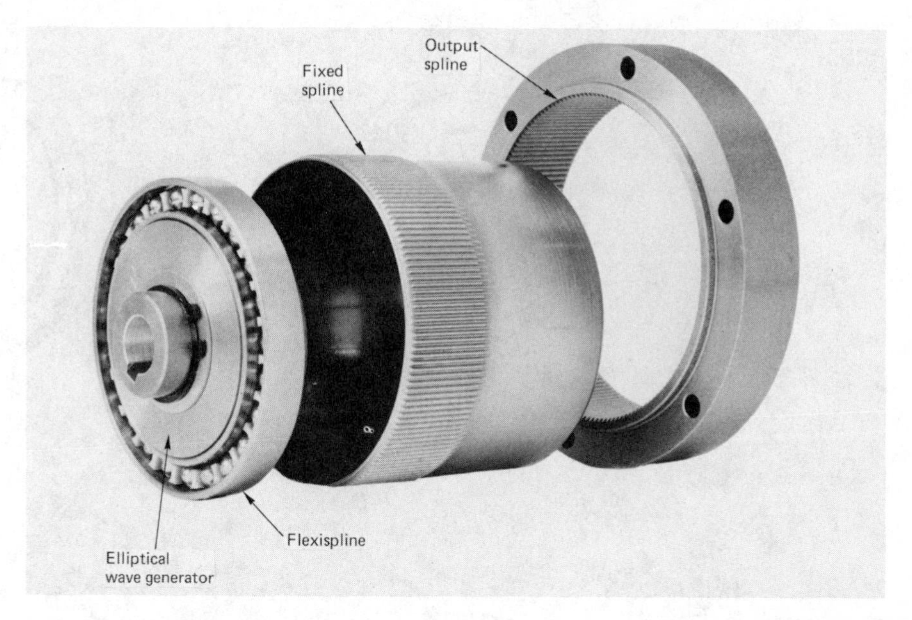

generator can still rotate within it. Since the wave generator is elliptical, it causes the flexible spline to 'bulge' as it rotates. The result is that the flexible spline is progressively forced to mesh with the internally cut teeth of the fixed spline. The fixed spline has two more teeth than the flexible spline, cut around its internal circumference. Thus, by the time the wave generator has rotated through one revolution, the flexible spline is two teeth out from where it started. For every revolution of the wave generator, there will be a relative movement of two teeth between the flexible spline and the fixed spline.

These conditions combine to provide a speed reduction (relative to the input wave generator) equal to half the number of teeth on the flexible spline, i.e. if there are 300 teeth cut on the flexible spline, the speed reduction will be 150:1. Since the fixed spline cannot move, positive transmission causes the flexible spline to rotate around the inside of the fixed spline. The output spline (which has the same number of teeth as the flexible spline) meshes with, and is driven by, the flexible spline at the reduced speed. Note that the action of the rotating wave generator causes the flexible spline to precess (rotate slowly) in the opposite direction to it. The high torque capability of the drive is a result of a high number of teeth (between 15% and 20%) being in contact (and transmitting power) at any one time.

Cyclo-drives The term 'cyclo' is a shortened form of the word 'cycloid' and is derived from the fact that the teeth on elements of the drive (the cycloid discs) are cycloidal in shape. (The cycloid shape is discussed in Chapter 2, Section 2.2/0.)

The cyclo-drive consists of four principal elements. With reference to Fig. 3/12 they are

1. The *eccentric input shaft* (high speed input).
2. The *cycloid disc* (one or more depending on the speed reduction).
3. The *outer rollers* which remain fixed.
4. The *output shaft* with drive pins and rollers (slow speed output).

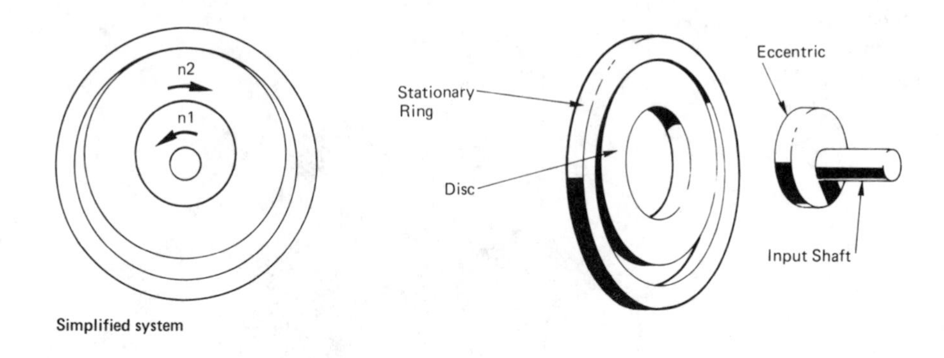

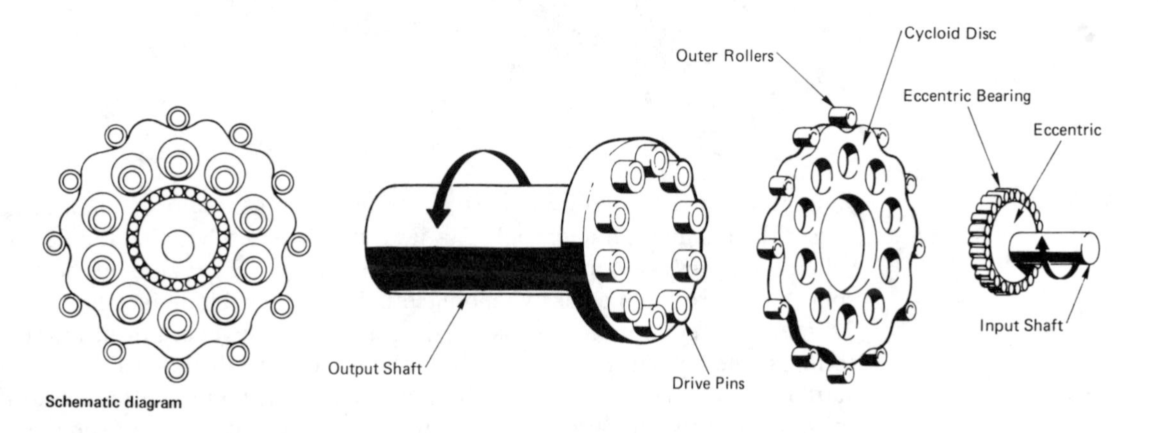

Fig. 3/12 Elements of a cyclo drive speed reducer [*courtesy: Centa Transmissions Ltd.*]

The eccentric input shaft, which rotates within rolling bearings, rotates within the cycloid disc. The action of the eccentric causes the cycloid ring to roll around the outer rollers, with a sort of bobbing action, in a reverse direction to that of the input shaft. Since there is one less tooth on the cycloid disc than there are outer rollers, the cycloid disc rotates by one tooth pitch per input shaft revolution. The reduction ratio is thus determined by the number of teeth on the cycloid disc. Rotation of the cycloid disc is transmitted to the output shaft by a number of drive pins and rollers projecting through holes in the cycloid disc. A single-disc system can achieve reduction ratios of between 6:1 and 87:1. In double-disc systems, using identical discs, these ratios may be squared. Multiple-disc configurations can achieve reductions of many thousands to one. Note that the direction of the output shaft will be opposite to that of the input shaft for a single-disc system, but in the same direction for a double-disc system. The use of rolling elements means that there is no sliding friction and the transmission is smooth, quiet and reliable. Cyclo gearboxes can be easily married with compact servomotors and braking systems to produce highly compact, efficient and effective axis drive systems for robotic applications.

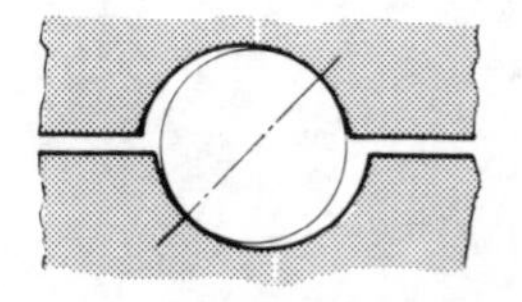

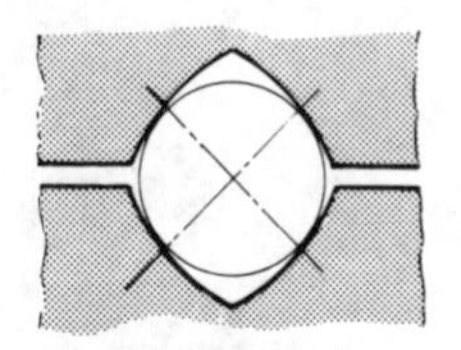

Fig. 3/13 Semi-circular and gothic arc leadscrew forms

Leadscrew drives A further common means of achieving speed reduction is by leadscrews. These have the additional feature that they convert rotary motion into linear motion. Acme-form screw threads (found on most conventional machine tools) are not suitable since they exhibit high-friction/high-wear characteristics, have poor power efficiency, and produce high lost motion due to excessive backlash. In addition, they are not accurate or repeatable enough for robotic applications, and they can only be used at relatively low operating speeds.

Modern **recirculating ball leadscrews** replace sliding friction with rolling friction. Both the leadscrew and the nut have a precision ground form into which ball bearings are allowed to run. The form is so designed that contact only takes place on opposing faces of the assembly. The geometry of the ground thread form can be either semicircular or a 'gothic arc' (ogival) form. Both these leadscrew thread forms are illustrated in Fig. 3/13. Enough ball bearings are inserted into the assembly to completely support the screw/nut assembly. Internal or external ball tracks allow the balls to continually circulate as the screw rotates, in the same manner as within a ball race.

Such leadscrews are very efficient and a load carried in a vertical plane is quite capable of sustaining a downward movement under its own weight. The rigidity of a drive system and the positioning precision can be increased by pre-loading the nut assembly. This is achieved by using two nuts and mounting them such that a pre-load exists between them. They may be pre-loaded in tension or compression. Pre-loading tends to slightly increase both wear and the torque required to drive the screw.

An alternative leadscrew design is the **planetary roller leadscrew**. This device employs precision threaded rollers which engage with a precision ground leadscrew. As the leadscrew is rotated, the rollers act like a planetary gearbox by rotating around the leadscrew within the nut assembly. The rollers have gear teeth cut into each end. These teeth then mesh with the internal teeth of a gear ring within the nut assembly. This prevents axial movement of the rollers but allows linear motion to be transmitted through the drive. High

Fig. 3/14 Recirculating ball leadscrews [*courtesy: PGM Ballscrews*]

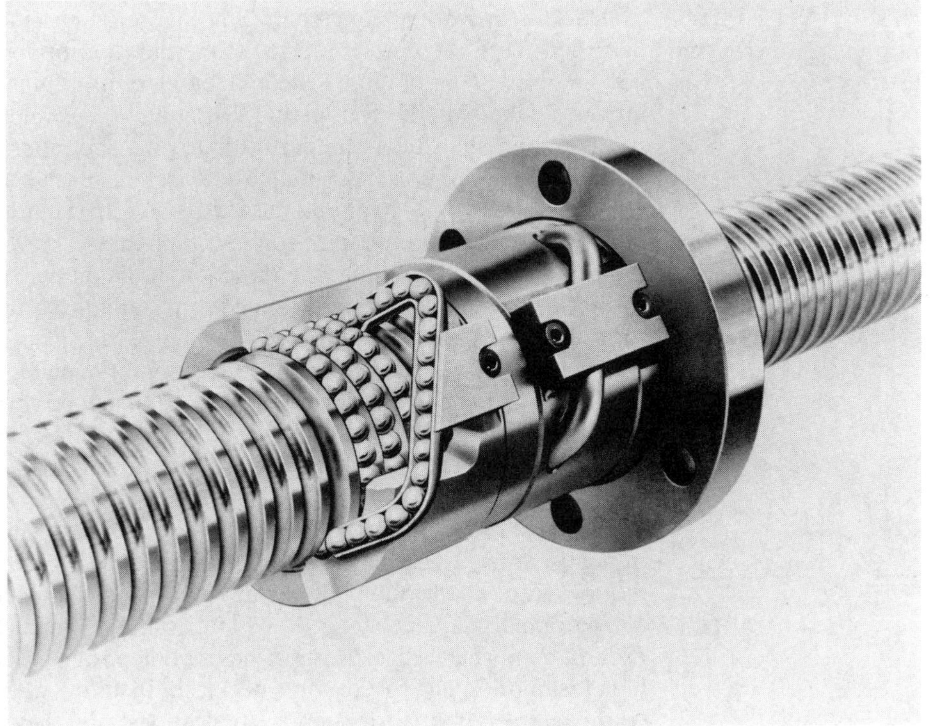

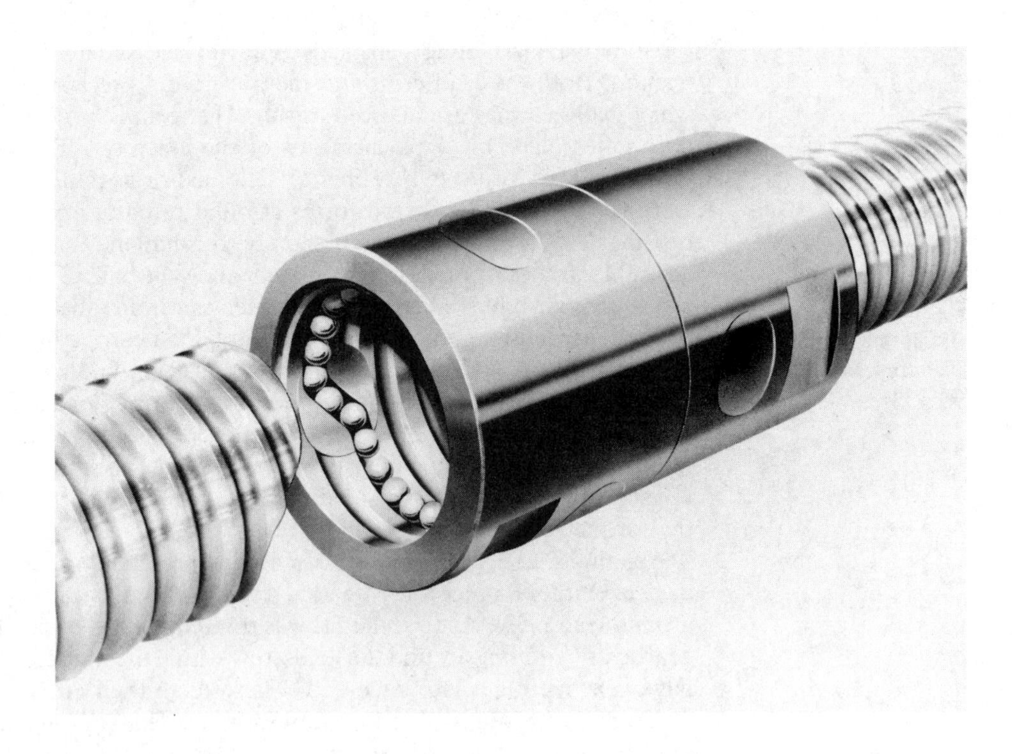

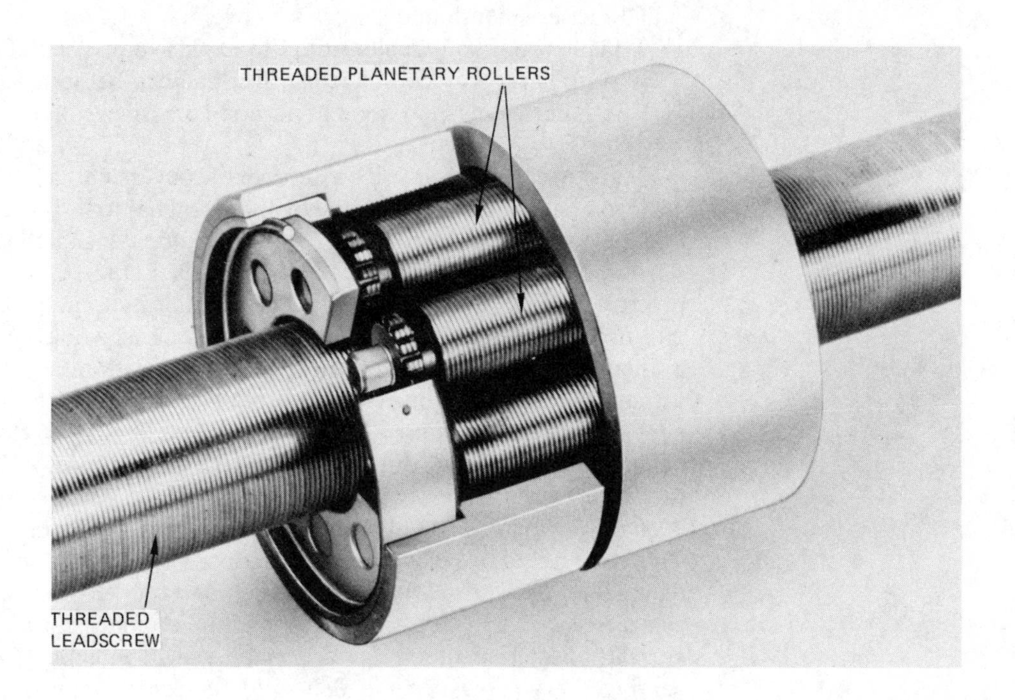

Fig. 3/15 Planetary roller leadscrew [*courtesy: ESE Ltd.*]

precision movement can be achieved since the precision leadscrew can be ground with a multi-start thread. The multiple contact points achieved by the precision threaded rollers offer heavy load carrying capability. Longer working life is achieved since the multiple contact points evenly distribute the applied loads. The threaded rollers are continuously in contact with both the nut and the leadscrew. There are no loose balls to recirculate and thus higher rotational speeds can be attained. Recirculating ball leadscrews are shown in Fig. 3/14 and a planetary roller leadscrew is shown in Fig. 3/15.

3.4/4 Hydraulic axis drive systems

Hydraulic axis drive systems operate by forcing pressurised hydraulic fluid (usually a mineral-based oil with chemical additives) around a closed circuit. The pressure generated can then be used to cause hydraulic motors to rotate or pistons to extend and retract. It is termed **fluid power**, a term which also covers pneumatic systems.

The claimed advantages for hydraulic systems include:

a) High efficiency and good power-to-weight ratio.
b) Suitable for high power applications.
c) Complete and accurate control over speed, position and direction of actuators.
d) Accuracy maintained under extreme load conditions, since hydraulic oil is virtually incompressible (0.5% at 7000 kN/m^2).
e) Minimises the number of mechanical linkages required.

f) Precise, smooth and shockless movement.
g) Flameproof and ideally suited to explosive or volatile environments.
h) Automatic protection against overload can be designed in.
i) Self-lubricating (low wear) and non-corrosive.

Fluid power systems involve a number of conversion stages in order to finally achieve axis motion. Initially, the hydraulic oil has to be pressurised, generally by an electric motor driving an hydraulic pump which delivers the hydraulic oil under pressure. Since the electric motor is the start of the conversion process, it is termed the *prime mover*. Earlier hydraulic systems often used internal combustion engines as the prime mover. The fluid power so generated is then converted into mechanical power via hydraulic cylinders or rotary actuators. Losses in conversion are small, usually manifesting themselves as heat, vibration and noise. Hydraulic systems require a closed circuit, and exhausted fluid must be directed back to the pump (by return lines into a sump tank) for recirculation.

Hydraulic pumps may deliver fixed quantities of oil per motor revolution. They are termed **fixed displacement pumps**. These pumps are acceptable when demand for pressurised fluid is relatively constant. As robot axes operate, they can often vary the demand on the hydraulic pump. To accommodate this, **variable displacement pumps** are employed. These pumps are capable of varying the volume of fluid delivered (at constant pressure) according to demand. The principles of operation of the common hydraulic pump types are illustrated in Fig. 3/16.

Fig. 3/16 Principles of operation of common hydraulic pumps

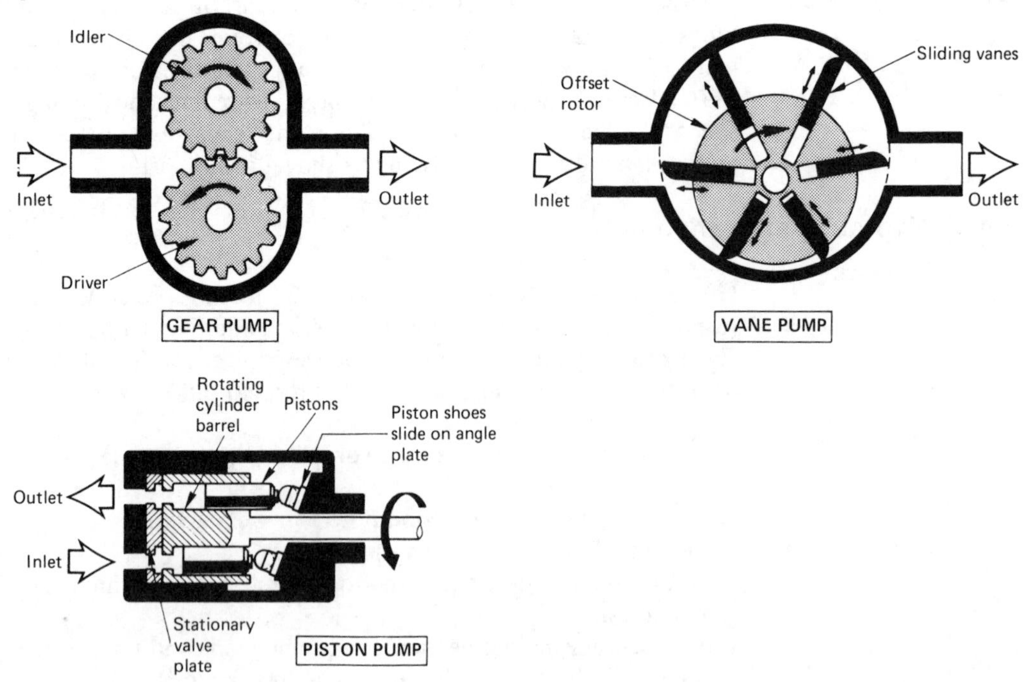

Hydraulic fluid systems may drive rotary vane-type actuators called **rotary actuators**, or be used to extend or retract pistons within cylinders. The latter are termed **linear actuators** and are employed on prismatic joints. Note that engineers refer to hydraulic cylinders rather than hydraulic pistons. The principles of the rotary and linear actuator are illustrated in Fig. 3/17.

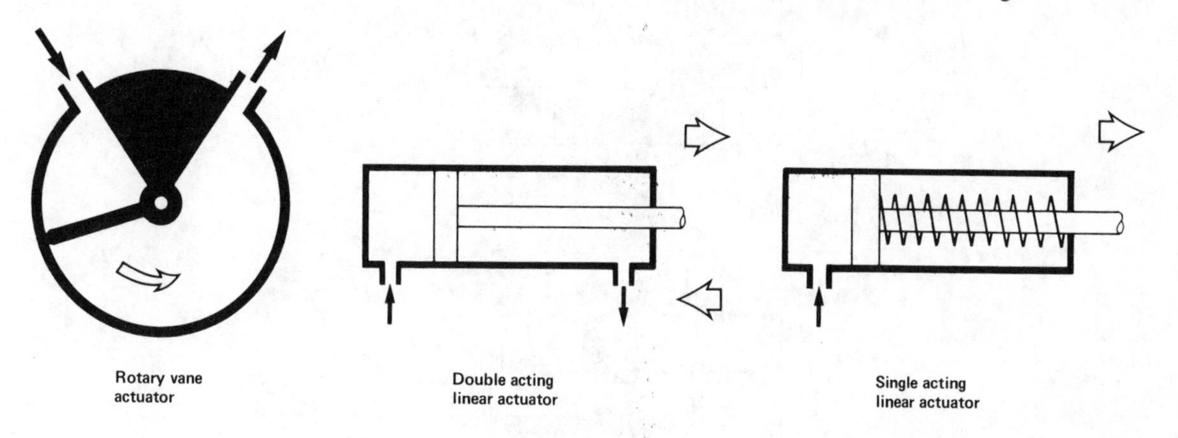

Fig. 3/17 Principle of rotary and linear actuators

The orifices in hydraulic components that admit hydraulic fluid, and allow it to be exhausted, are termed **ports**. The ports may act as inputs or outputs for the hydraulic fluid depending in which direction the fluid is routed. This allows both rotary and linear actuators to be powered in both forward and reverse directions.

The rotary actuator comprises a shaft joined onto a rotary vane that separates two ported chambers. As hydraulic fluid is displaced from one chamber (by pressurised fluid entering the other), the vane rotates and drives an output shaft.

The linear actuator works on the same principle. Some cylinders can only be extended under fluid power. Retraction of the piston is accomplished by a compression-type return spring accommodated within the cylinder body. These are termed **single acting cylinders**. For more precise control, and power availability in both forward and reverse directions, there are **double acting cylinders**. These cylinder types are illustrated in Fig. 3/17. They are available in various standard sizes, in a number of designs and offering a number of different mounting arrangements.

The control of pressure, flow and direction of hydraulic fluid within a hydraulic circuit is accomplished by discrete hydraulic valves. The most common type is the **spool valve**. A precision ground shaft (the *spool*) moves horizontally within the valve body. In so doing, it either covers or uncovers inlet or outlet ports, thus allowing the passage of fluid. Spool valves may be operated in a number of ways: manually, or by air, electrical or small hydraulic signals. The latter are termed **pilot operated valves**, since only a relatively small signal (the *pilot signal*) is required to actuate them. Control valves differ in the number of ports they have, the number of ways in which they allow fluid to pass, and their method of actuation. The principle of operation of a typical spool valve is illustrated in Fig. 3/18. Hydraulic circuits are connected via rigid (steel) pipes or by flexible (reinforced) hoses.

Fig. 3/18 Principle of operation of a hydraulic spool valve

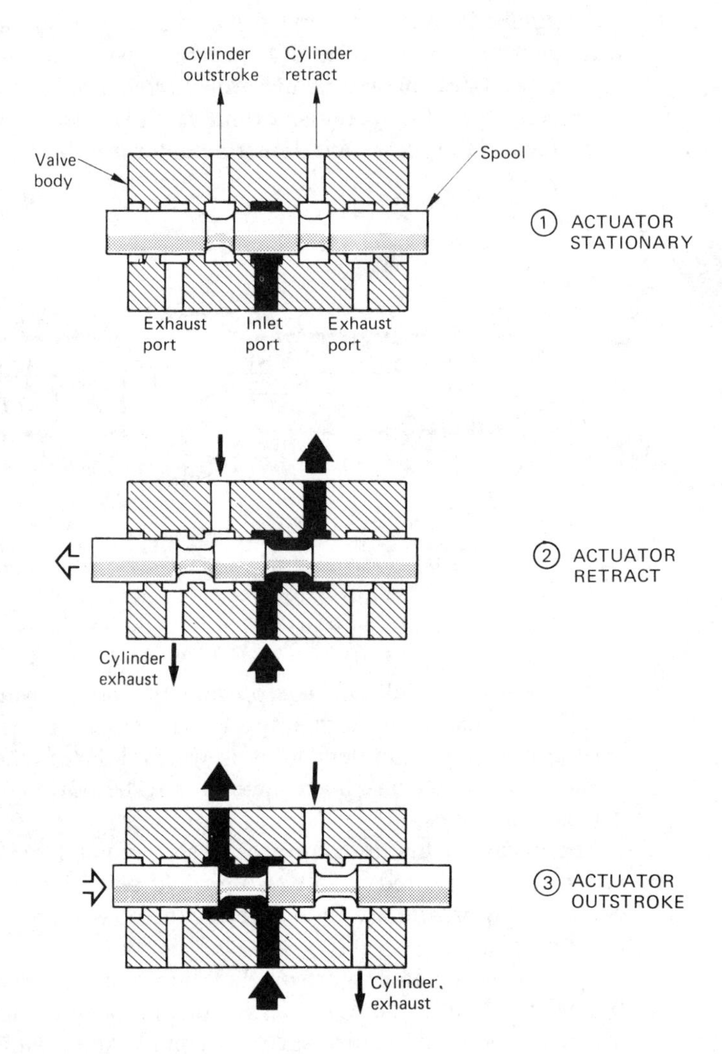

Hydraulic fluid must be carefully conditioned before it is allowed into a hydraulic circuit. It must first of all be filtered of any dirt and foreign particles. Such particles could prevent the correct operation of precision components with unsafe consequences, or contribute to premature component wear or damage. All air within the system must be similarly removed. The presence of air can cause compressibility of the fluid and also lead to a condition called *cavitation*, which can cause extreme wear and damage to circuit elements if allowed to persist. All leaking joints, as well as reducing the efficiency of the circuit, are also potential entry points for the ingress of air and dirt. The efficient sealing of hydraulic cylinders is a constant problem and the cause of much associated long-term maintenance. Cooling of the fluid may also need to be carried out to maintain an acceptable working temperature of the hydraulic fluid.

3.4/5 Pneumatic axis drive systems

Pneumatic fluid power systems use compressed air as the transmission medium. Many of the principles involved in hydraulic system operation also apply to pneumatic system operation. Since air is compressible, precise control of speed and position is difficult to achieve and power application is less than that achievable by hydraulic systems. However, the compressibility of air can also be used to advantage in absorbing shock loads, preventing damage due to overload, and providing a compliance ('give') that may be required in many practical applications. Claimed advantages of pneumatic systems include the following:

a) Air is plentiful and compressed air is readily available in most factories.
b) Compressed air can be stored and conveyed easily over long distances.
c) Compressed air need not be returned to a sump tank; it can be vented to atmosphere after it has performed its useful work (although exhausts may be irritatingly loud).
d) Compressed air is clean, explosion-proof and insensitive to temperature fluctuations, thus lending itself to many applications.
e) Operation can be fast and speeds and forces can be infinitely adjusted between their operational limits.
f) Digital and logic switching can be performed by pneumatic fluid logic elements.
g) Pneumatic elements are simple and reliable in construction and operation and are relatively cheap.
h) Pneumatic cylinders, comprising proximity sensors to sense the position of the piston, allow the easy integration of pneumatic systems with computer sensing and control.

As with hydraulic systems, a number of stages of conversion have to be carried out. Compressors driven by a prime mover draw in and compress air from the atmosphere. They may be either piston compressors or turbine compressors. Both types are available in a number of designs. **Piston compressors** operate by reducing the volume of air and increasing its pressure. **Turbine compressors** operate by drawing in air, compressing it by mass acceleration, thus converting kinetic energy into pressure energy. Compressors are specified according to their delivery volume and their delivery pressure. Air delivered by the compressor can be fed directly into the circuit (after conditioning), or accumulated in a pressure vessel called a *receiver*. The receiver, as well as acting as a storage container, also provides a reservoir for sudden demands in circuit pressure.

Before compressed air is admitted into a circuit it must first be conditioned to ensure reliability of circuit operation. Conditioning, carried out by special-purpose *service units*, performs three principal functions. First, all dirt and foreign particles are filtered out. This increases system reliability by lessening wear and damage, the tendency to jam or block up delicate control devices, and the mis-operation of circuit components. Secondly, since air absorbs moisture, which can condense out within the circuit components, it has to be dried. Finally, the air is enriched with a fine oil mist to provide lubrication for the various system components.

Compressed air can be applied to both linear and rotary actuators. Linear actuators, in the form of air cylinders, are similar, in construction and principle of operation, to their hydraulic counterparts. Rotary actuators, commonly termed **air motors**, consist of a rotary vane driven by the applied air pressure. They are clean, simple in operation, relatively cheap, and provide inherent cooling. They are usually geared by high reduction gear units to maintain smooth operation.

Pneumatic systems have the added advantage that logic switching elements can be implemented with low-pressure air supplies.

3.4/6 Control of axis position

The ideal means of measuring axis position would be to continuously measure the position of the end effector relative to the programmed or robot datum. Under such conditions, no losses would be incurred. Unfortunately, this has not yet been achieved. Measurement from the end effector is made almost impossible by their different designs, the cargoes, components and tools they carry, and obstructions through which a programmable movement may sweep. Positional feedback is provided by measuring axis movement indirectly via compact and discrete measuring transducers attached to both fixed and moving robot members.

Basically, two types of **position measuring transducer** are employed on industrial robots:

a) Angular transducers
b) Linear transducers

Such transducers must be capable of measuring position (from a prescribed datum) or distance moved (incremental or absolute) from some reference point. Each controlled axis requires a position measuring transducer.

Position measuring transducers may be either analog or digital in their design. An analog quantity is one that can vary continuously, between limits, over a period of time. Voltage, temperature, sound, etc. are analog quantities. By contrast, digital devices operate between two (and only two) distinct states. These two states may be referred to in a variety of ways: high/low, on/off, 1/0, set/reset, mark/space. There can be *no* in-between state. In digital systems, the 'high' state is represented by the presence of a voltage and the 'low' state by the absence of a voltage. By convention, these states are most commonly represented by 1 and 0 respectively.

3.4/7 Angular position measuring transducers

Angular transducers operate by measuring angular rotation, normally of an axis leadscrew or rotary joint. The most popular angular transducers are discussed below.

SINGLE RADIAL GRATING In this system a translucent disc is ruled with a series of radial lines (**gratings**). The resulting disc is made up of alternate (uniform) transparent and opaque areas. This disc is then keyed to the axis leadscrew. A collimated (parallel) light source and photocell arrangement is

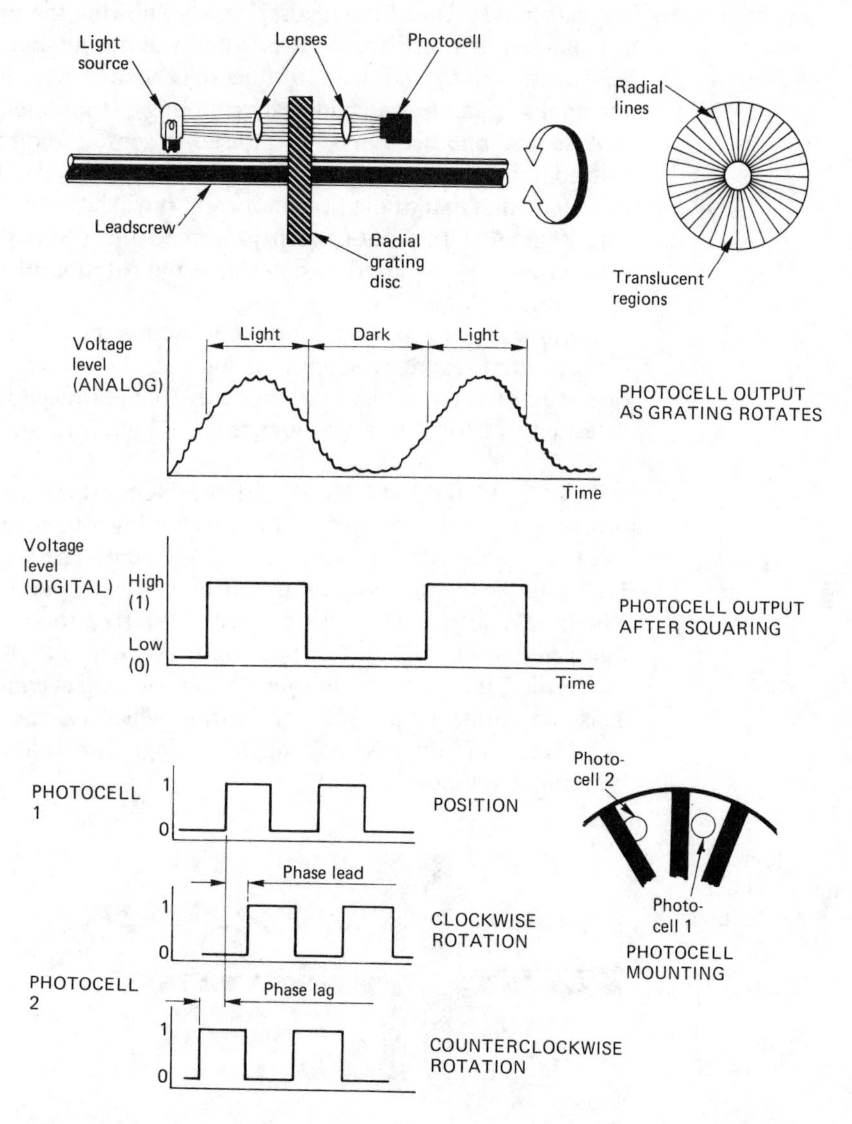

Fig. 3/19 Principle of operation of a single radial grating

mounted so that, with rotation of the leadscrew, the photocell will sense alternate light and dark areas. The set-up is shown in Fig. 3/19*a*.

As a dark area of the disc is gradually uncovered, the light intensity falling on the photocell gradually builds up until it reaches a maximum when it is completely uncovered. As the disc continues to rotate, the following dark area starts to impede the light intensity falling on the photocell. The light intensity will gradually reduce until it is zero, when the dark area again cuts out light transmission. Since the photocell gives out a voltage that is proportional to the intensity of the light it receives, the resulting output resembles the shape of a sine wave. An electrical component known as a Schmitt Trigger converts this sinusoidal-shaped output into more of a square-shaped (or pulsed) output. This is illustrated in Fig. 3/19*b*.

The output can now be recognised as a series of discrete pulses, each pulse corresponding to a transparent region of the disc. Each pulse represents an angular movement of the leadscrew. By knowing the number of lines (hence the number of transparent areas) engraved on the disc, and (in the case of a leadscrew drive) the lead of the axis leadscrew, the movement of the manipulator can be calculated by counting the number of pulses sensed.

A second photocell must be utilised to sense the direction of rotation. By positioning the second photocell as shown in Fig. 3/19*c*, the pulsed output will be identical to that of the first photocell. It will however be slightly out of step. The degree of this out-of-step is termed the **phase difference**. The phase difference can be sensed to determine the rotation of the leadscrew. This is illustrated in Fig. 3/19*c*.

Accuracy of position measurement by this method is limited by two main factors. First, there is a physical limit to the number of lines that can be engraved on a given disc, and secondly, the photocell itself requires a gap of a certain size for it to sense variations in light intensity.

RADIAL MOIRÉ FRINGE GRATINGS Two radial gratings can be positioned adjacent to each other. One grating is fixed and the second is keyed to (and rotates with) the rotating axis element. The arrangement is shown in Fig. 3/20*a*. If the disc centres are offset, a moving pattern of lines is produced when the shaft rotates. This pattern is referred to as a Moiré Fringe and is illustrated in Fig. 3/20*b*. Observation shows that if there are *n* lines engraved on the disc, this pattern will rotate *n* times during each revolution of the shaft. This is a form of optical magnification which can be sensed by the use of photocells as in the case of a single grating. Direction of rotation is detected by using a second photocell.

Fig. 3/20 Principle of operation of radial Moiré fringe gratings

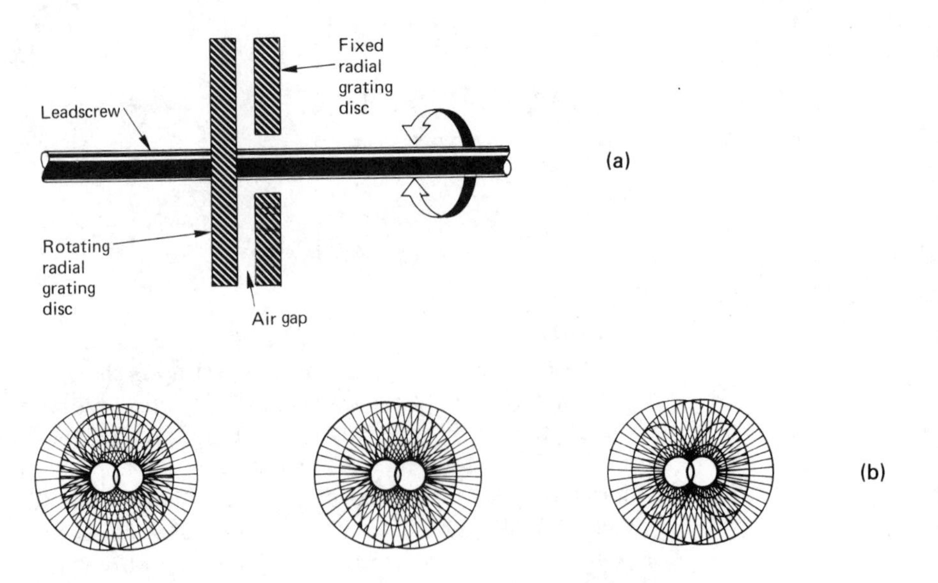

SHAFT ENCODER In this arrangement, a disc is encoded by etching transparent and opaque (light and dark) regions in the form of a binary pattern. (See Chapter 4 for a detailed treatment of the binary system.) Essentially, at any angular position of the encoder disc, a series of photocells transmit a binary code depending on the form of the light and dark areas sensed. The binary code transmitted is an absolute indication of the position of the leadscrew throughout its rotation. A shaft encoder is illustrated in Fig. 3/21. The accuracy of the encoder disc is a function of how many photocells (binary digits) are used to make up the binary code. Although three are shown in Fig. 3/21, up to twelve may be employed. Using twelve binary digits in this way can produce a resolution of 1/4096 of a revolution.

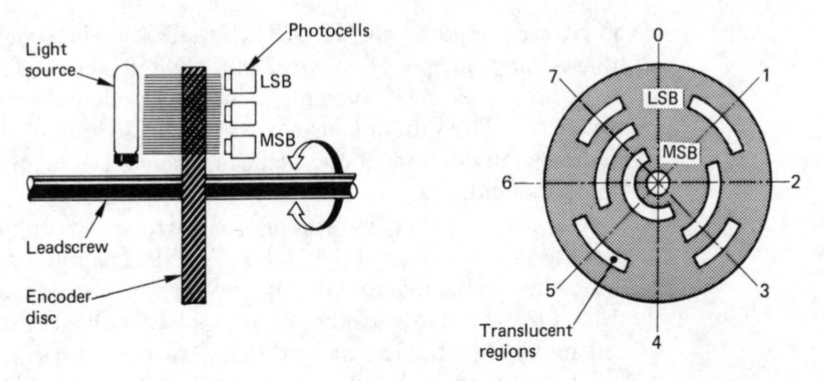

Fig. 3/21 A 3-bit shaft encoder

The binary code at each of the eight radial positions (corresponding to the decimal numbers represented at those positions), shown in Fig. 3/21, is generated by a unique combination of photocell ONs and OFFs. The resulting code is shown in the table below (1 represents ON and 0 represents OFF by convention). In practice, malfunctions can occur if the photocells become skewed from the radial line. At the transition points, between (decimal) numbers 1 and 2, 3 and 4, and 5 and 6, a number of photocells change their states at once. This can be another source of malfunction. This condition becomes more common as the number of binary digits (photocells) increase. To overcome this problem, the natural binary code is modified so that, at any transition point, a change in only one binary digit is required. The resulting code is known as the **Gray code**. Gray coded discs increase reliability but require the use of additional decode circuitry within the transducer.

Decimal Number	*Natural Binary Code*	*Gray Code*
0	000	000
1	001	001
2	010	011
3	011	010
4	100	110
5	101	111
6	110	101
7	111	100

Each of the above transducers can indicate either the incremental or absolute amount of rotation within each revolution. The means of indication is repeated for each subsequent revolution. There is usually some further means to count the number of full revolutions turned through. In its simplest form, this can consist of a single photocell sensing through a single slot cut in the outside of the disc. Each pulse delivered by this photocell indicates one full revolution of the axis motor.

SYNCHRO and SYNCHRO RESOLVER The synchro analog transducer utilises the principle of *magnetic induction*, which is as follows. If a conductor carries an electrical current, a magnetic field is produced around that conductor. The situation also exists in reverse in that, if a conductor is moved in the vicinity of a magnetic field (or vice versa), a current will be induced in that conductor.

A series of stationary windings are arranged around the periphery of the synchro. This is termed the **stator**. A central spindle also carrying a winding is attached to the manipulator axis and rotates within the stator. This is termed the **rotor**. The stator windings are supplied with a constant AC voltage which sets up a magnetic field around them. As the rotor rotates, with a leadscrew, a voltage is induced in the rotor winding. This induced voltage varies from a minimum to a maximum, depending on its position relative to the stator windings, in a sinusoidal fashion. The magnitude (size) of this voltage represents the angular position of the rotor. By counting the number of complete revolutions and sensing the angular position of the rotor, via the induced voltage, an indication of rotary position is obtained. The principle of the synchro is illustrated in Fig. 3/22.

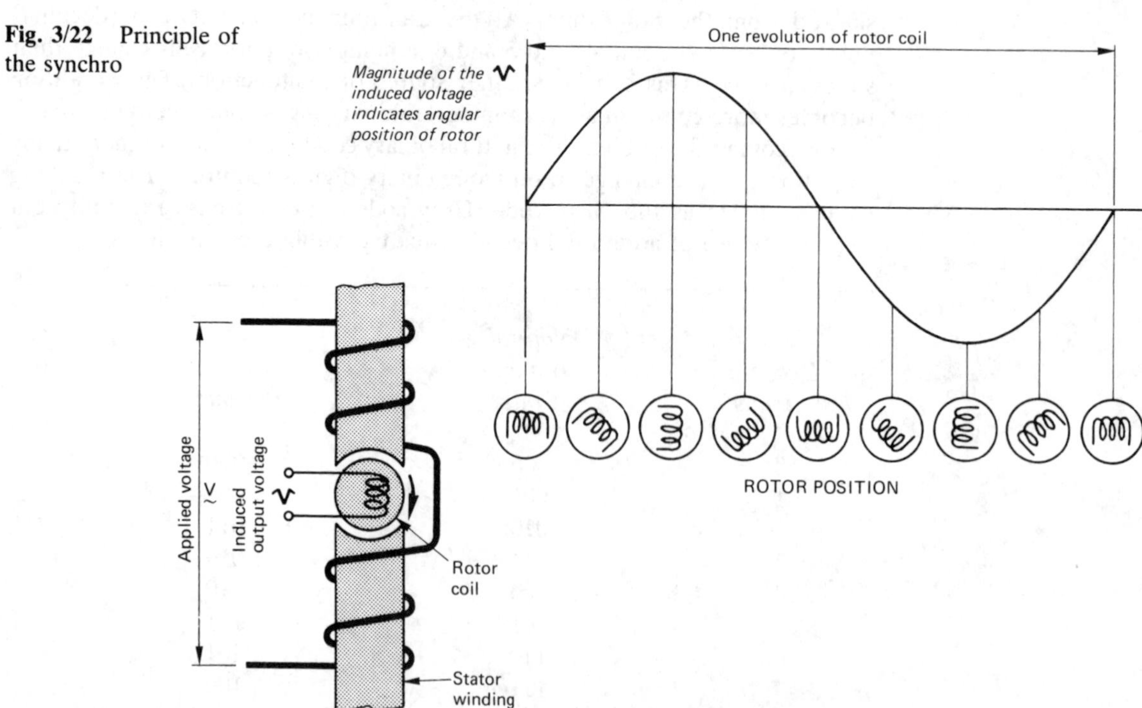

Fig. 3/22 Principle of the synchro

A synchro resolver works on a similar principle but has two stator windings positioned at right angles to each other. It is so called since it 'resolves' a voltage into two components (the *sine* and the *cosine*) of the angle made by the rotor. The windings are fed with an identical AC voltage but shifted in phase (out of step) by 90°. The induced voltage in the rotor winding will itself have a phase shift, with respect to the stator windings, which is determined by its angular position. In a synchro resolver it is the phase shift of the induced voltage that is proportional to the angular position of the rotor.

Angular position measuring transducers can be built discretely into, and form part of, the revolute joints of appropriate robots. As such they are well protected from damage and the industrial environment within which the robot is employed.

3.4/8 Linear position measuring transducers

Position measurement by angular transducers is ideal for robots that contain revolute joints. Linear measuring transducers operate by recording actual movement of the robot's prismatic joints. Errors due to backlash, torsional wind-up and pitch errors in leadscrew drives, for example, are eliminated. They have the disadvantage that they must be physically protected.

LINEAR GRATING A linear grating works on the same principle as the single radial grating. A precision linear scale engraved with close-spaced parallel lines is fixed to the moving arm. A light source and photocell detector are fixed on a convenient stationary element of the robot. This is illustrated in Fig. 3/23. As the transparent regions of the linear grating expose the light source, a pulse is registered by the photocell. By knowing the pitch of the engraved lines on the linear grating, and counting the number of pulses, the movement of the arm can be established. As with radial gratings, linear gratings utilise a second photocell to detect direction of movement.

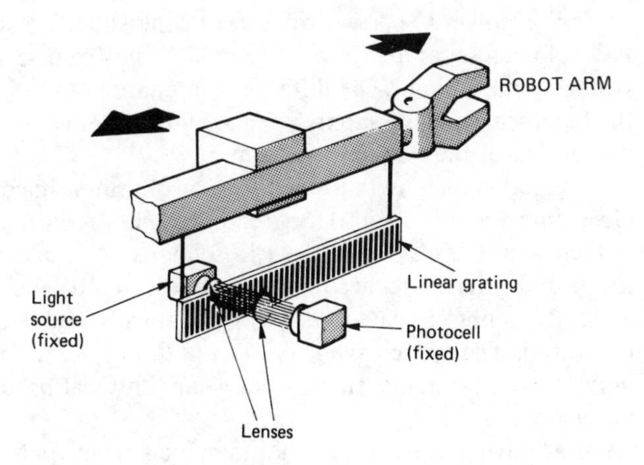

Fig. 3/23 Principle of operation of a single linear grating

Linear gratings can either be etched onto glass, in which case light is detected through them, or engraved onto stainless steel tapes, in which case light is reflected from them.

LINEAR MOIRÉ FRINGE GRATINGS Linear gratings may be limited in resolution for some applications. This is because of the physical limitation of the number of parallel lines that can be engraved, per unit length, to preserve a sufficient gap as required by the photocell.

Linear Moiré fringe gratings, working on the same principle as the radial Moiré fringe grating, can be arranged in a linear fashion. A fixed-scale grating is attached to the robot structure. A smaller index grating is affixed to, and moves with, the moving arm. The lines on the index grating are at the same pitch as those on the fixed grating, but are inclined at a slight angle. When the two gratings move relative to each other, the characteristic Moiré fringe pattern is observed to move across the grating. The intensity of the light falling on the photocell through the gratings varies sinusoidally as movement proceeds. A number of photocells can be arranged to monitor the same fringe movement to increase resolution. The order in which the photocells are excited (monitored by phase difference) determines the direction of motion of the manipulator arm.

Moiré fringe gratings can easily be reproduced to examine the effect. Simply rule some closely-spaced parallel lines on a sheet of paper. Reproduce this on a sheet of transparent film and experiment by placing the film over the paper, at a slight angle, and moving it from right to left. The fringe can be observed to move across the grating. Notice that reversing the relative direction of movement causes the fringe to move in the opposite direction.

INDUCTOSYN The inductosyn operates on the synchro-resolver principle but is effectively 'laid out flat'. It comprises a long fixed scale carrying a pattern of wires that repeat at regular intervals (every 2 mm or 3 mm). This performs the same function as the rotor winding in the synchro resolver. The winding is usually printed onto a glass scale. Glass is an insulator, is very stable, and has a linear coefficient of expansion. A second, smaller scale, carrying two similar patterns of repeating windings, is mounted as a slider on the moving arm. The slider performs the same function as the stator windings in the synchro resolver. The two windings of the slider are supplied with identical AC voltages separated in phase by 90°. The fixed-scale winding will have a voltage induced in it. The difference in phase between the induced voltage of the fixed scale and the supply voltage of the slider is a measure of the positional movement of the manipulator arm.

A big advantage is that one cycle of output voltage can easily be divided electronically (up to 1/500) for greater resolution. For practical purposes, most inductosyn scales are produced in sections. A number of sections are then fitted and aligned to accommodate the full axis movement. Alignment of windings is not crucial since it is magnetically coupled voltage that is being measured. The inductosyn is illustrated in Fig. 3/24. Since there is no contact between the elements, there is no wear. Physical protection must, however, be provided.

To achieve feedback, the output signals from each analog-type positional transducer need to be converted into a form that the control unit can accept. Since robot control units are digital devices, this form must be digital. Additional electronic circuitry is thus required to compare command signals and feedback signals. Such electronics employ standard devices such as

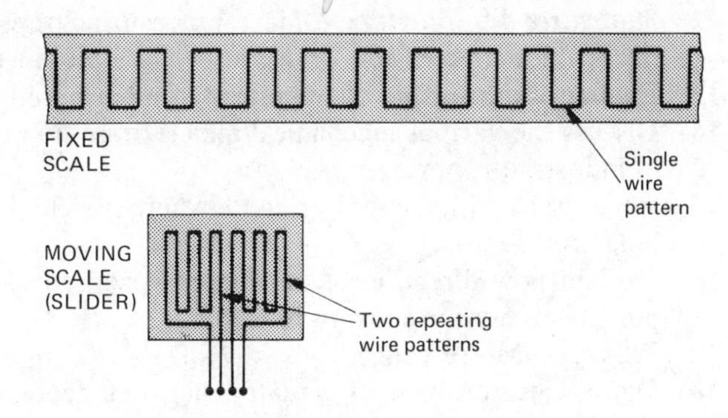

Fig. 3/24 Components of the inductosyn position measuring transducer

analog-to-digital convertors (ADC) or *phase-to-digital convertors* (PDC). The resulting output can then also be easily displayed in digital form for axis readouts.

All of the above transducers are indirect in that they do not indicate the precise actual position of the end effector in relation to the established working datum. Slight losses or errors are therefore inherent. Many efforts are made to reduce such losses to an absolute minimum. For example, inappropriate location of the measuring transducers contribute to likely losses within the total control system. Linear transducers should be mounted near to sliding surfaces and leadscrews, observing the need for accessibility for maintenance purposes. Losses in angular measuring transducers will be minimised if they are mounted at the free end of any leadscrews rather than at the driving end.

Questions 3

1 What are block diagrams and why are they so useful?
2 Explain, using block diagrams, the terms 'open loop' and 'closed loop' in the context of industrial robot control systems.
3 Explain what is meant by the term 'feedback', and state *two* types of feedback required in a closed loop robot control system.
4 Define the term 'hunting' and explain why it can occur.
5 Briefly explain the differences between 'point-to-point' and 'contouring' control in the context of industrial robot control systems.
6 What is the difference between linear and circular interpolation, and when would each be employed?
7 Define the terms 'repeatability', 'accuracy', 'resolution' and 'response' in the context of industrial robot control systems.
8 Draw a neat block diagram of a typical industrial robot system. Make brief notes on each of the elements of the system.
9 Discuss the advantages and disadvantages of electrical drives as an actuation system for use in industrial robots.
10 Compare and contrast the use of hydraulic and pneumatic actuation systems for use in industrial robots.
11 Briefly explain the difference in operation of a 'shaft encoder' and a 'radial grating' transducer used for positional feedback.

12 State *three* different types of linear measuring transducer that can be used for positional feedback, and briefly explain their principles of operation.

13 Explain the principle of operation of a harmonic drive.

14 Discuss the various mechanical means of motion transmission employed in industrial robots.

15 What is a 'tachogenerator' and where would it be employed on an industrial robot?

16 Explain how direction of movement is sensed when using grating-type positional measuring transducers.

17 What is a Moiré Fringe?

18 Define the term 'error-actuated negative feedback' in the context of closed loop control system operation and explain how its design can lead to a condition known as 'hunting'.

19 Explain why external sensors may be employed in industrial robotic situations.

Computer Considerations 4

4.1 Computer systems

4.1/0 What is a computer?

Modern computers are devices which are configured to manipulate information (**data**) according to a set of rules in a very precise and ordered manner. The set of rules by which the computer operates is called a **program**. Much of the flexibility of computer systems is derived from the fact that different programs for handling the same data (from a common database) can be designed, and run on a single computer system, or that a single program may work on many different sets of data. *All* the actions the computer takes are predetermined by the programs it runs; it cannot (as yet) display 'intelligence' of its own.

The principal advantages of computer systems may be summarised thus:

a) They can work at incredible speeds.
b) They can work consistently without requiring rest, supervision or other services.
c) They are extremely accurate.
d) They do not forget.
e) They can issue warnings and reminders based on the most up-to-date information.
f) They are compact and relatively cheap.

These factors translated into industrial terms mean greater productivity, lower costs, consistency of output, and tighter controls on the day-to-day running of the organisation.

4.1/1 Computer sub-systems

A computer system can be designed to be a general-purpose data-handling system. For example, a typical home or business computer may display immense versatility using colour and graphics. Alternatively, it may be designed to carry out a very specific task. Such a computer system is then known as a **dedicated computer**. Dedicated computers are found in washing machines, cash registers, petrol pumps, etc., and in the machine control units of industrial robots.

No matter what their intended purpose, all computer systems have a common structure. The concept of block diagrams was introduced in Chapter 3 (Section 3.1/1). Using the same approach, a block diagram of a typical computer system is illustrated in Fig. 4/1.

Fig. 4/1 Block diagram of a typical computer system

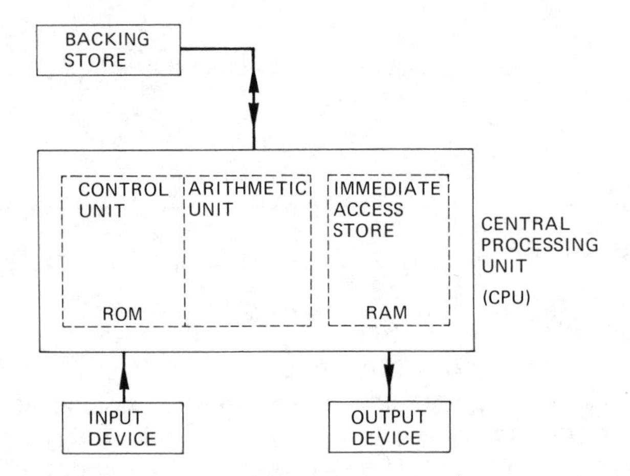

The system comprises a **central processing unit** or CPU which communicates with various **peripheral devices**. The CPU is itself made up of three distinct sub-sections:

a) The **control unit** is responsible for coordinating all the functions carried out by the computer. These functions include processing the current program instructions in the correct order at the correct time, and also looking after the various 'housekeeping' tasks of loading in and saving programs, handling error conditions, etc.
b) The **arithmetic unit** is responsible for any calculations but can also select, sort and compare information which can be used to execute some parts of the program in preference to others.
c) The **immediate access store** is the **internal memory** of the computer used to store the program and results of any calculations performed by the arithmetic unit. Its operation is electronic and, since there are no moving parts, data can be accessed in nanoseconds (one thousand millionths of a second).

In a robot control system, the **input devices** are numerous. First, the move sequence program must be entered into the control system memory. This may be done via the teach pendant, a paper tape reader, a magnetic tape or disc reader, or a host computer. Secondly, the control system also monitors the status of the elements of the robot itself in terms of axis speed, axis positions, etc., through various transducers. Similarly, the control system also has many **output devices** which it must address: the system display, magnetic tape/disc units, as well as axis motors, end effectors, position measuring transducers, etc.

The wide variation in peripheral devices which need to communicate with the controller inevitably leads to some incompatibilities in the form in which each device accepts and transmits its information. To overcome this, it may be necessary to provide an **interface** between the control unit and the various peripheral devices. An interface is a collection of electronic circuitry designed to make information in one form compatible with information in another form.

4.1/2 What is a microprocessor?

A **microprocessor** is basically the 'brain' of a digital computer system and is equivalent to the combined Control Unit and the Arithmetic Unit. It is a complex of micro circuits housed in a single **integrated circuit** or *chip*. A typical microprocessor can obey about 70 separate simple instructions but these may be combined, and used in different modes, to form powerful programs. Since each instruction can be executed in a mere 2 to 10 microseconds (2–10 millionths of a second) complex tasks can be performed very rapidly. An indication of the size and speed of the microprocessor is given in Fig. 4/2.

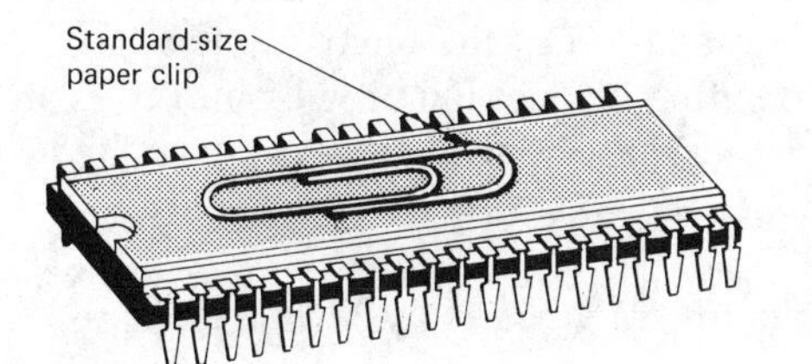

Fig. 4/2 An indication of the size and speed of a microprocessor

The flexibility in configuring and programming a microprocessor to perform practically any function attracts system designers in diverse fields. The ability to change the function of a piece of equipment by changing some instructions in the microprocessor system instead of redesigning a printed circuit board has obvious attractions. Thus, industrial robot control system suppliers are able to offer control system updates, at nominal cost, as extra facilities become available.

It is the use of the microprocessor that gives the microcomputer its name and makes the application of compact, low-cost computing power possible. However, a microprocessor is of little use on its own and must be combined with some form of input, some form of output, and, in most cases, extra memory, to form a usable computer system.

4.1/3 Computer memory

A computer is capable of doing only one thing at a time. The result of a calculation, for example, must be stored away in the computer's memory before the next program instruction is carried out, otherwise it will be lost.

Computer memory can be visualised as rows of storage **locations** or 'pigeon holes'. Each storage location is capable of storing a single **character** of information. Characters may be alphabetic letters, numbers or punctuation marks (termed *alphanumeric* characters).

The action of storing (or saving) a character in a memory location is termed **writing to** memory, and the action of fetching a character from a memory location is termed **reading from** memory. When a character is read from a memory location its contents remain unchanged. Thus, data may be read any number of times. This is why a move sequence held in memory may be executed any number of times without reloading it. When a character is written into a memory location, it will overwrite the original contents of that memory location. This makes it possible to edit robot move sequences held in the memory of the control unit and thus make minor modifications to the program, at the machine.

The smallest unit of memory is one character or *one byte* of information. Memory capacity is usually quoted in terms of **kilobytes** or K for short. In the ISO number system we understand the prefix kilo- (or k) to represent multiples of 1000, e.g. 1 kilogramme = 1000 grammes, 1 kilometre = 1000 metres, etc. In computer terms, however, kilo- represents multiples of 1024. The reason is as follows. The ISO system is a decimal-based system (it counts with a base of 10) and 1000 is a round multiple of 10, i.e. 10^2. Computer systems work in binary (which count with a base of 2) and successive multiplication by 2 cannot 'hit' 1000 exactly. The nearest multiple is 1024 or 2^{10}. (See Section 4.2 for a further explanation of the binary system.)

Computer memory devices fall broadly into two categories. Reference back to Fig. 4/1 shows that the control unit is also labelled ROM and the immediate access store labelled RAM:

> **ROM** stands for **Read Only Memory**. As its name implies, it can only be read by the processor. This type of memory has its contents built (or burnt) into it when it is manufactured. It is thus used to house the dedicated main control program used by the control unit. The contents of ROM cannot be erased even by removing the power supply. It is said to be a **non-volatile** memory device.
>
> **RAM** stands for **Random Access Memory**. This is memory that can either be written to, or read from, at random. It is also termed read/write memory. The same memory locations can be used over and over again to hold different programs and data. Some robot control systems allow more than one sequence program to reside in memory at the same time with the facility to switch between them. The contents of RAM are wiped out if the power supply is removed; it is thus termed a **volatile** memory device.

Typical robot control systems have between 16K and 64K of RAM.

Another popular and very useful memory device is the **EPROM**. This stands for *Erasable Programmable Read Only Memory*. It is in fact a re-usable ROM. It can be programmed using a device called an *eprom programmer*. Once programmed, the EPROM acts likes a ROM in that its contents are non-volatile. An EPROM can be erased, if required, by subjecting it to ultra-violet light for between 20 and 30 minutes. It can then be re-programmed. These devices make control system program updates extremely simple and cost effective.

Backing store refers to external storage such as magnetic tape or disc devices. Backing stores are necessary to provide permanent and back-up copies of move sequence programs for future use. Intelligent utilisation of backing stores can create easily accessible libraries of often-used move sequence programs, subroutines or macros. Backing store devices will be discussed more fully in Section 4.4/2.

The general term **hardware** describes all the physical parts of a computer/robotic system. **Software** refers to computer or move sequence programs and the media on which they are stored, and **firmware** describes computer programs stored on a chip, such as the dedicated main control program held, in ROM, within the robot control unit.

4.2 The binary system and its importance

4.2/0 Number bases

In everyday arithmetic we make use of numbers expressed in the decimal (or denary) system. Decimal means 'ten', which is called the *base*. A **number base** is simply the number of digits, including zero, which the system can use. Because zero must always be included, the actual digits used go up to one less than the base itself. Thus, in the decimal system the digits used are 0, 1, 2, 3, 4, 5, 6, 7, 8 and 9.

Very large or very small numbers are expressed using the same digits but in different positions. For example, the number 3256 means:

3-thousands + 2-hundreds + 5-tens + 6-ones (or units)

More precisely, each number is separated, in value, from its neighbour by a power of 10 (the base). The digit having the least value is at the right-hand end.

1000s	100s	10s	Units	position value
10^3	10^2	10^1	10^0	number base value
3	2	5	6	decimal value

Many other number systems are in use for specialist applications but they all have the common features outlined above, i.e. the use of a base and a value related to position which is a power of the base.

4.2/1 Binary numbers

A computer is an electronic device and as such relies on levels of voltage for its operation. If the computer were constructed to work according to the decimal system, then it follows that it would require ten distinct voltage levels or states, one for each digit. In practice this is difficult to implement and it is far easier for electronic engineers to create electronic circuits using only two states. The two states then become: a voltage being present or no voltage being present; or an electronic switch being open or an electronic switch being closed. For this reason, the binary number system (with a base of 2) using only two digits, 0 and 1, is ideal for computer applications.

Because there can be no 'in-between' state using this system, it is referred to as being *digital*. The electronics employed is known as **digital electronics** and the computers that result are termed **digital computers**. Robot control systems employ digital computers.

Since only two digits are used, binary numbers take on the form of long strings of 1s and 0s, for example 101011. Each digit in the string is known as a **bit**. This comes from a contraction of the term BInary digiT. To understand the significance of such numbers we must revert to the theory of number bases introduced above. As with other number bases, the digit at the extreme right has the least value. The rightmost digit in a binary number is thus known as the **least significant bit** (LSB). Conversely, the leftmost digit, having the highest value, is termed the **most significant bit** (MSB).

Using the above analysis, the binary number 101011 can be translated as follows:

32s	16s	8s	4s	2s	1s	position value
2^5	2^4	2^3	2^2	2^1	2^0	number base value
1 MSB	0	1	0	1	1 LSB	binary value

Note how the value of the bits increases by a power of 2 as we proceed leftwards. In fact the value of each bit is *double* that of its predecessor.

Since long strings of 1s and 0s can be somewhat confusing and difficult to evaluate, it is often more convenient for us to translate the binary number into its *decimal number equivalent*. From the above **bit pattern**, the binary number consists of:

$$(1 \times 32) + (1 \times 8) + (1 \times 2) + (1 \times 1) = 43$$

Thus, the decimal equivalent of 101011 is 43.

Exercise Calculate the decimal equivalent of the binary number 1101011 and the binary representation of 77.

4.3 Digital code systems

4.3/0 Numerical information

There appears to be two apparent drawbacks of the binary system as far as robot programming and control are concerned. The first is that as the value of the number increases so does the length of the bit pattern. The length of the bit pattern is known as the **word length**. Thus, large dimensions require longer word lengths than small dimensions:

43	=	01011	6-bit word length
143	=	10001111	8-bit word length
943	=	1110101111	10-bit word length

Secondly, there seems no way of representing fractional numbers. In the decimal system we have the decimal point. Digits to the left of the decimal point successively increase by a power of ten and numbers to the right of the decimal point successively decrease by a power of 10, e.g. 256.83.

Both drawbacks are overcome by using a system known as **binary coded decimal** (BCD). In this system each *digit* in the decimal number is given its own *binary word*. The separate words are then transmitted one after the other in the correct (decimal) order. For example, the number 256.83 expressed in BCD reads:

1st word	0010	2
2nd word	0101	5
3rd word	0110	6
		.
4th word	1000	8
5th word	0011	3

	bit 4	bit 3	bit 2	bit 1
1	0	0	0	1
2	0	0	1	0
3	0	0	1	1
4	0	1	0	0
5	0	1	0	1
6	0	1	1	0
7	0	1	1	1
8	1	0	0	0
9	1	0	0	1
10	1	0	1	0
11	1	0	1	1
12	1	1	0	0
13	1	1	0	1
14	1	1	1	0
15	1	1	1	1
16	0	0	0	0

Fig. 4/3 Combinations of 4 bits can represent 16 individual codes

The bytes of information are stored consecutively in memory or on an appropriate storage media. One convenient storage media is punched tape, since it is possible to see the data as it is stored. It is a simple step to imagine the representation above being reproduced as rows of hole patterns in a punched tape. Where there is a binary 1, punch a hole; where there is a binary 0, do not punch a hole. The presence or absence of a hole can then be sensed by a variety of means and then decoded, according to the same principles, by computer logic within the robot controller. Similarly, the individual bits can be represented on magnetic media as different audible frequencies, or directional magnetic fields.

Binary coded decimal has added advantages especially when using punched (or magnetic) tape as a program medium. Word lengths, and hence tape widths, can be kept to realistic sizes. Note how it is possible to represent all the digits, 0–9, using a word length of only four bits. This presentation makes it easier to recognise hole patterns for the digits 0–9. This enables operators to 'read' the numbers, and hence the dimensions, punched in the tape.

4.3/1 Alphanumeric characters

It was stated earlier that a memory location is capable of holding a single character of information, and that this character can be a digit, an alphabetic character or a punctuation symbol. We have just seen how numbers can be represented using binary coded decimal but what of the *alpha* characters?

If we examine the system of representing numbers in binary in the previous section a little more closely, it becomes apparent that each digit is merely a unique arrangement of 1s and 0s, i.e. a code! The numerical value assigned to any particular bit pattern just happens to coincide with the weighting associated with each position, starting from the right-hand side. Indeed, any arrangement of 1s and 0s can be chosen to represent any digit so long as the device **decoding** the number uses the same interpretation as the device **encoding** the number.

We used four bits to represent ten digits (0–9). In fact, the combinations of four bits allows the representation of sixteen different symbols. Fig. 4/3 illustrates the possible combinations. It follows, therefore, that alpha characters and punctuation symbols could also be represented in coded form, by a unique bit pattern of 1s and 0s.

4.3/2 The ASCII code

The most widely used coding system for computer applications is the **ASCII** (pronounced ASKEY) code. This is an American-devised code and the letters stand for *American Standard Code for Information Interchange*. It is this coding system on which the ISO recommendations for CNC (computer numerical control) codes is based (BS3635 Part 1 : 1972). Robot control programs are a form of CNC control. Indeed some robot manufacturers adhere to the same ISO coding system as used on the majority of CNC metal cutting machine tools.

The ASCII code represents alphanumeric characters using a 7-bit word length. It is interesting why a word length of 7-bits was chosen. For computing applications it is necessary to represent, at least, the following characters:

26	upper case (capital) letters
26	lower case (small) letters
10	numeric digits (0–9)
4	arithmetic symbols (+, −, ×, ÷)
1	decimal point
67	characters

The possible number of unique combinations that can be achieved using a 6-bit binary word length is 64 (or 2^6). Clearly, this is inadequate for even the minimum **character set** suggested above. A 7-bit word length is thus used which can accommodate many more characters: for example, a full range of punctuation marks and many commonly used symbols such as brackets, pound and dollar signs, percent (%) and ampersand (&) symbols, etc., with 'spare capacity' for special *control codes* used for sending instructions to peripheral devices.

Exercise Calculate how many characters, in all, can be represented using 7-bits.

Since computers are now commonly communicating with CNC machine tools and industrial robots directly, it demonstrates sound judgement to base the CNC character codes on the established 'standard' for the computer industry.

4.3/3 The ISO 7-bit numerical control code

BS3635 Part 1: 1972 specifies the standard **ISO 7-bit code** that is recommended for CNC applications by the International Standards Organisation. In fact, this code is a sub-set of the ASCII code, comprising some 50 characters. It will be appreciated that the full ASCII character set would not be appropriate, in its entirety, for CNC of industrial machines and robots. For example, there will be no need to specify lower case letters and many textual symbols for such a specialised application.

The full ISO character set, binary representation and decimal equivalents are shown in Fig. 4/4.

Also shown is the ISO code set representation as it appears when punched in paper tape. Punched paper tape represents the single most popular storage medium for CNC machining applications but is less popular for robotic applications. It is, however, very convenient for examining the make-up of

Fig. 4/4 Representation of the ISO character set

DECIMAL EQUIVALENT	BINARY REPRESENTATION							NAME OF CHARACTER	CHARACTER SYMBOL	REPRESENTATION IN PUNCHED TAPE								
										P	7	6	5	4	F	3	2	1
	b_7	b_6	b_5	b_4	b_3	b_2	b_1				b_7	b_6	b_5	b_4		b_3	b_2	b_1
0	0	0	0	0	0	0	0	NULL	NUL						•			
8	0	0	0	1	0	0	0	BACKSPACE	BS	•				•	•			
9	0	0	0	1	0	0	1	TABULATE	TAB					•	•			•
10	0	0	0	1	0	1	0	END OF BLOCK	LF					•	•		•	
13	0	0	0	1	1	0	1	CARRIAGE RETURN	CR	•				•	•	•		•
32	0	1	0	0	0	0	0	SPACE	SP	•		•			•			
37	0	1	0	0	1	0	1	PROGRAM START	%	•		•			•	•		•
40	0	1	0	1	0	0	0	CONTROL OUT	(			•		•	•			
41	0	1	0	1	0	0	1	CONTROL IN	)	•		•		•	•			•
43	0	1	0	1	0	1	1	PLUS SIGN	+			•		•	•		•	•
45	0	1	0	1	1	0	1	MINUS SIGN	–			•		•	•	•		•
47	0	1	0	1	1	1	1	OPTIONAL BLOCK SKIP	/	•		•		•	•	•	•	•
48	0	1	1	0	0	0	0		Ø			•	•		•			
49	0	1	1	0	0	0	1		1	•		•	•		•			•
50	0	1	1	0	0	1	0		2	•		•	•		•		•	
51	0	1	1	0	0	1	1		3			•	•		•		•	•
52	0	1	1	0	1	0	0		4	•		•	•		•	•		
53	0	1	1	0	1	0	1		5			•	•		•	•		•
54	0	1	1	0	1	1	0		6			•	•		•	•	•	
55	0	1	1	0	1	1	1		7	•		•	•		•	•	•	•
56	0	1	1	1	0	0	0		8	•		•	•	•	•			
57	0	1	1	1	0	0	1		9			•	•	•	•			
58	0	1	1	1	0	1	0	ALIGNMENT FUNCTION	:			•	•	•	•		•	
65	1	0	0	0	0	0	1		A		•				•			•
66	1	0	0	0	0	1	0		B		•				•		•	
67	1	0	0	0	0	1	1		C	•	•				•		•	•
68	1	0	0	0	1	0	0		D		•				•	•		
69	1	0	0	0	1	0	1		E	•	•				•	•		•
70	1	0	0	0	1	1	0		F	•	•				•	•	•	
71	1	0	0	0	1	1	1		G		•				•	•	•	•
72	1	0	0	1	0	0	0		H		•			•	•			
73	1	0	0	1	0	0	1		I	•	•			•	•			•
74	1	0	0	1	0	1	0		J	•	•			•	•		•	
75	1	0	0	1	0	1	1		K		•			•	•		•	•
76	1	0	0	1	1	0	0		L	•	•			•	•	•		
77	1	0	0	1	1	0	1		M		•			•	•	•		•
78	1	0	0	1	1	1	0		N		•			•	•	•	•	
79	1	0	0	1	1	1	1		O	•	•			•	•	•	•	•
80	1	0	1	0	0	0	0		P		•		•		•			
81	1	0	1	0	0	0	1		Q	•	•		•		•			•
82	1	0	1	0	0	1	0		R	•	•		•		•		•	
83	1	0	1	0	0	1	1		S		•		•		•		•	•
84	1	0	1	0	1	0	0		T	•	•		•		•	•		
85	1	0	1	0	1	0	1		U		•		•		•	•		•
86	1	0	1	0	1	1	0		V		•		•		•	•	•	
87	1	0	1	0	1	1	1		W	•	•		•		•	•	•	•
88	1	0	1	1	0	0	0		X	•	•		•	•	•			
89	1	0	1	1	0	0	1		Y		•		•	•	•			•
90	1	0	1	1	0	1	0		Z		•		•	•	•		•	
127	1	1	1	1	1	1	1	DELETE	DEL	•	•	•	•	•	•	•	•	•

the coding system. These principles of data representation remain the same in both electronic and magnetic storage media.

We shall use the representation in paper tape of Fig. 4/4, for convenience, to examine the make-up of the coding system. Remember, the presence of a hole represents a binary 1 and the absence of a hole represents a binary 0. We shall consider the vertical columns as *tracks* or *channels* on the tape.

This figure warrants close scrutiny since, although the code primarily has a 7-bit word length, it is represented, in punched tape, by 8 bits. The most significant bit (the leftmost track) is included, and reserved for, an error-checking device called a **parity check**. This is a system for detecting certain errors that could be present during data transmission and does not form part of the coding system itself. It is a system applied primarily to data stored, and transmitted, via tape media. (Parity and error checking will be discussed in Section 4.4/4.)

Let us first consider how the digits 0–9 are represented. Remember, we must ignore track 8, the parity track, since this does not form part of the coding system. We must also ignore the vertical stream of *feed holes* that run the entire length of the tape, in between tracks 3 and 4. These are used to physically transport the tape through tape reading and tape punching devices.

First, *all* the digits have a hole punched in tracks 5 and 6. From then on, the digits can be read as decimal equivalents of their respective binary punchings. Thus, the digit 1 is equivalent to binary 1 plus holes in tracks 5 and 6. The digit 2 is equivalent to binary 2 plus holes in tracks 5 and 6, digit 3 is equivalent to binary 3 plus holes in tracks 5 and 6, and so on.

The alphabetic characters follow a similar system. *All* the alphabetic characters A–Z have a hole punched in track 7. From then on the letters follow an ascending binary count from 1 to 26. For example, the letter A is equivalent to binary 1 plus a hole in track 7. The letter B is equivalent to binary 2 plus a hole in track 7, the letter C is equivalent to binary 3 plus a hole in track 7, right the way up to the letter Z which is equivalent to binary 26 plus a hole in track 7.

The DELETE character has holes punched in *all* tracks since this represents the only way of nullifying every character.

This leaves only 12 characters from the full 50 that perhaps do not lend themselves to easy recognition. It is easy to see why punched tape has retained its popularity when over 75% of the characters can so easily be read by the human operator.

You may have noticed the absence of the decimal point within the ISO character set. The recommendations of BS3635 (Section 6.3/1) state that the decimal point shall *not* be shown in a control tape and that its position shall be implied by the format of the dimension. For example, the control system may be designed to assume a decimal point after reading the first three or four digits of an input position dimension. In such systems the digits representing the dimensional positional data may have to be 'padded out' with leading zeros where necessary. For example, to input a dimensional move of 9.86 mm into such a system would require the dimension to be expressed as 00986. Where control systems do not require this 'padding', they are said to employ **leading zero suppression**. However, the control unit of some systems and other data input means, for example a teach pendant or a host computer, may require the decimal point character.

The coding system discussed above is valid no matter what form of data input is employed—it is merely convenient to use the medium of punched tape to explain its make-up. Particular collections of coded bytes taken together may be recognised by the control system as *instructions* that will cause particular actions to be initiated. Other collections of coded bytes may simply represent numerical or positional *data*.

Note that the word byte, used to describe the smallest unit of memory required to store a single character of information, is borrowed from the computer scientists. More formally, 1 byte consists of 8 bits. So each single memory location within the computer is itself made up from eight bits. Each bit is then capable of being *set* (binary 1) or *reset* (binary 0) to form the unique bit patterns of the individual characters.

4.3/4 Inputs and outputs

Move sequence programs held in memory are collections of coded bytes stored sequentially in consecutive memory locations. Programs entered via a teach pendant are similarly converted into coded bytes and also stored sequentially in memory. The control unit and the main operating system control program decide where, in memory, the incoming bytes will be stored as part of their 'housekeeping' tasks. When the stored move sequence program is executed, the control system moves to the beginning of the stored program memory and interrogates each of the program bytes in turn. It differentiates between instructions and data and acts accordingly. This simplified description can easily be visualised since memory organisation is conveniently ordered into individual memory locations, and program execution is strictly sequenced within the computerised control unit.

However, much data and instruction transfer must also occur between the control unit and devices external to the control unit. These external devices are not necessarily under the full control of the robot control system, and may operate using completely different voltage standards. Such information transfer may be **input** (into the control system), or **output** (out from the control system). The devices involved in the exchange are termed **input devices** and **output devices** respectively. Typical input devices include positional feedback transducers, limit switches, external sensors and so on. Typical output devices include axis drive elements, end effectors, conditionally operated peripheral devices, status and display indicators, etc.

All input to, and output from, the control unit must be in digital form. If devices give out, or require, signals in other forms then additional circuitry must arrange this. Two common forms of conversion involve changing analog signals into digital signals (**analog-to-digital conversion**), and digital signals into analog signals (**digital-to-analog conversion**). Both forms of conversion require readily available components built into purpose-designed interfaces. Both these types of conversion will be more fully discussed in Chapter 6, Section 6.2. To outline the principle by which input and output take place between the computerised control unit and external devices, we shall assume the use of digital signals only. Normal digital voltage levels (called **logic voltages**) operate by using 0 volts to represent logic 0 and 5 volts to represent logic 1. In practice, any voltage between 0 V and 0.8 V is treated as logic 0, and any voltage greater than 2.4 V is treated as logic 1 (the 'in-between' states cannot guarantee correct operation of the devices). This enables reliable circuit operation whilst allowing reasonable tolerances between individual circuit components. The voltages may also be referred to as **TTL levels**.

We have already seen that an individual memory location comprises eight bits. Imagine that, electronically, each individual bit of a designated memory location is connected to the pin of an electrical connector. It can be arranged that, when a particular bit is set to logic 1, the pin to which it is connected is energised with 5 V. Similarly, when an individual bit of the memory location is set to logic 0, the pin to which it is connected is set to 0 V. The memory location is often termed a **data register**. It can be seen that to energise any combination of the eight pins of the connector with a signal of 5 V simply requires a single byte to be written into the data register. This can easily be done under program control and illustrates how external devices can be controlled. The bits can be used individually (acting as simple switches), or in combination with each other, depending on the application. Input from external devices may be accomplished in the same way. A separate dedicated input data register may be provided or a single dual-purpose register may be used for both functions.

Fig. 4/5 Concept of Data Registers and Data Direction registers to control input and output of digital signals.

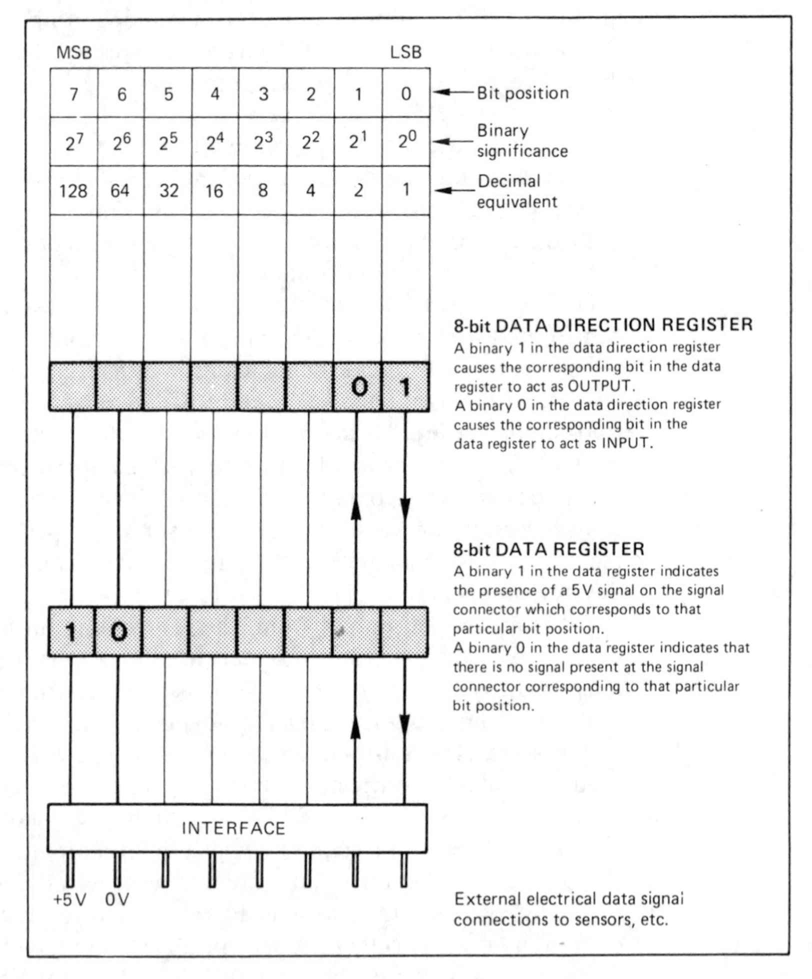

When a single register is used, there must be some way of differentiating between an input and an output signal, which will prevent two signals clashing on the same I/O line. One common approach is illustrated in Fig. 4/5. A second designated memory location, called a **data direction register** is arranged to work in conjunction with an associated data register. The individual bits of the data direction register control the direction of the data signal on corresponding bits of the data register. For example, a bit set to binary 1 in the data direction register may cause the corresponding bit of the data register to operate as output. This means that the data register can only be written to and cannot be read. Conversely, a bit set to binary 0 in the data direction register will cause the corresponding bit of the data register to operate as input. In this case the data register may be read, but writing to it will have no effect. The states set by the data direction register are mutually exclusive in that only one condition can be set at a time. This prevents mis-operation due to misinterpretation of the signals present at the data register and also the possibility of data clashes. An industrial robot control system will typically support 16 addressable input/output bits.

The organisation of memory incorporates some means of **synchronising** input and output signals (using separate control signals), and some means of maintaining the voltage levels at the connector (called a *latch*). Separate counters (or *timers*) may also be incorporated to provide timed delays. Collectively, all these features make up a *port*, in this case an **input/output port**. They are available within special integrated circuits called **port chips**, **versatile interface adaptors** (VIAs) or **programmable input/output** chips (PIOs), and provide a cheap and convenient means of circuit design and construction. All connections between the computerised control unit and external devices should be protected from reverse or excess voltages. Excess heat and the application (accidental or otherwise) of excess voltage remain common causes of mis-operation or damage to electronic circuits and components.

4.3/5 Logic operations

Arithmetical operations (all based on the operations of addition and subtraction) carried out on data bytes form the basis on which numerical calculations are performed. Logic operations are a second type of manipulation carried out on bytes of data. They are useful when dealing with individual bits within a byte. One family of logic operations is called **combinational logic** or **gating** operations. These operations may be performed by hardware (using *logic gates* housed within integrated circuits), or by software (using *logical operators*) from within a control program language. We shall discuss the operation and application of the two most common logical operations AND and OR by considering their hardware implementation as logic gates.

The **logic gate** operates by comparing two or more input signals to the gate, in a particular way. The output signal from the gate depends upon the type of logic gate (which fixes the rules by which the comparison should be made), and the state of the input signals. Consider, for simplicity, that there are two input signals to a logic gate. Each of these signals, at any point in time, may be either logic 1 or logic 0. According to the rules of binary, this means that there can be four possible input combinations (i.e. 2^2) to the logic gate. They are:

	Input A	*Input B*
combination 1	0	0
combination 2	0	1
combination 3	1	0
combination 4	1	1

The output from the logic gate, for each of the input combinations, is determined by the type of comparison or logic gate used. The results may be tabulated, as above, in the form of a **truth table**.

An **AND gate** gives a logic 1 output if, and only if, both inputs are at logic 1. Any other combination results in an output of logic 0. The truth table for the AND operation is therefore:

Input Signals *A*	Input Signals *B*	Output Signal *A* AND *B*
0	0	0
0	1	0
1	0	0
1	1	1

Consider an application where a robot loads a component into a machine tool. For safe operation two conditions must be met:

a) The machine spindle must be stationary.
b) The machine must be empty.

Since both conditions must be met before the loading operation is initiated, they may be combined using the AND condition. It is a simple matter to provide sensors that can sense the two conditions. The output from the sensors can then be used as the inputs to a two-input AND gate. The output signal from the AND gate can be the signal that initiates the robot load cycle, conditional on both its inputs being satisfied.

An **OR gate** gives a logic 1 output when one OR other of its inputs is at logic 1. An **Inclusive OR gate** (simply OR) also gives an output of logic 1 if both inputs are at logic 1. An **Exclusive OR gate** (EOR) gives an output of logic 0 if both inputs are at logic 1, thus excluding the ambiguous AND condition of the input signals. The truth tables for both types of OR operation are:

INCLUSIVE OR

Input Signals *A*	Input Signals *B*	Output Signal *A* OR *B*
0	0	0
0	1	1
1	0	1
1	1	1

EXCLUSIVE OR

Input Signals *A*	Input Signals *B*	Output Signal *A* EOR *B*
0	0	0
0	1	1
1	0	1
1	1	0

Consider an operation where a robot unloads components from a conveyor and places them on pallets. When a pallet is not available, such as when a fully laden pallet is being transported away, the robot still unloads the conveyor and places the components into a buffer store. If a pallet is present, the robot load sequence may then be initiated when there are components present on the conveyor OR in the buffer store. Sensors can be used to detect the presence of components at both the conveyor and the buffer store. The outputs from the sensors can be fed into the inputs of an OR gate and the resulting output signal used to initiate the palletising sequence. By monitoring the individual input lines of the OR gate, the robot can be directed to pick components from either the conveyor or the buffer store. It would be a simple matter to further refine the sequencing to include the AND condition of a pallet being present. By combining individual input signals through a variety of logic gates, complex situations can easily be modelled.

Such situations may be treated mathematically by using techniques known as **Boolean Algebra**. Logic gates are available with multiple inputs but the principles remain the same. Typical logic gates are shown in Fig. 4/6.

Fig. 4/6 Logic gates symbols and their implementation within integrated circuits

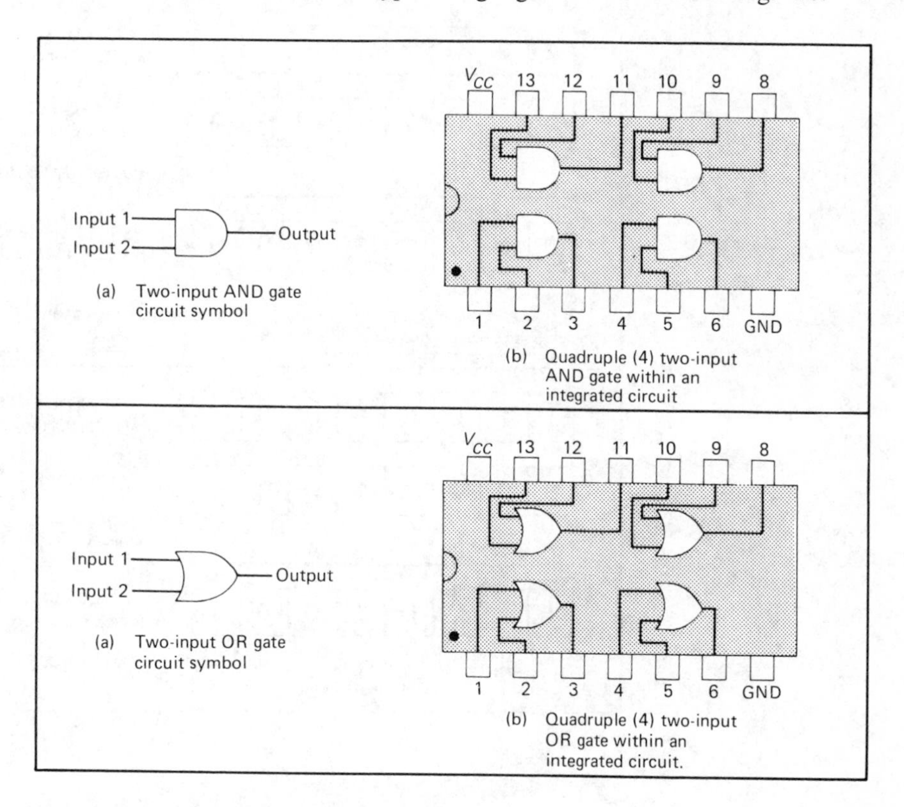

Software logical operations are performed in exactly the same way but they operate simultaneously on every bit within a particular byte. Consider that a particular I/O port has been set up so that it receives inputs from eight different sensors. When the sensors are active, a logical 1 appears at the corresponding bit position of the I/O port. The inputs from the various sensors can be read from the port, at any particular moment in time, as a single byte of data. Logical operations can be used to decode the byte of information to detect whether a particular sensor is active or not. For example, suppose that we wanted to ascertain whether the sensor feeding bit 4 was active. The decimal equivalent of bit 4 is $2^3 = 8$. A software instruction equivalent to 'Port contents AND 8' will yield a positive (non-zero) result if the sensor is active, and a zero result if the sensor is not active. This true or false (binary) outcome is easily sensed by the program logic. The reasoning is illustrated in Fig. 4/7*a*.

Fig. 4/7 Process of masking to condition input and output signals

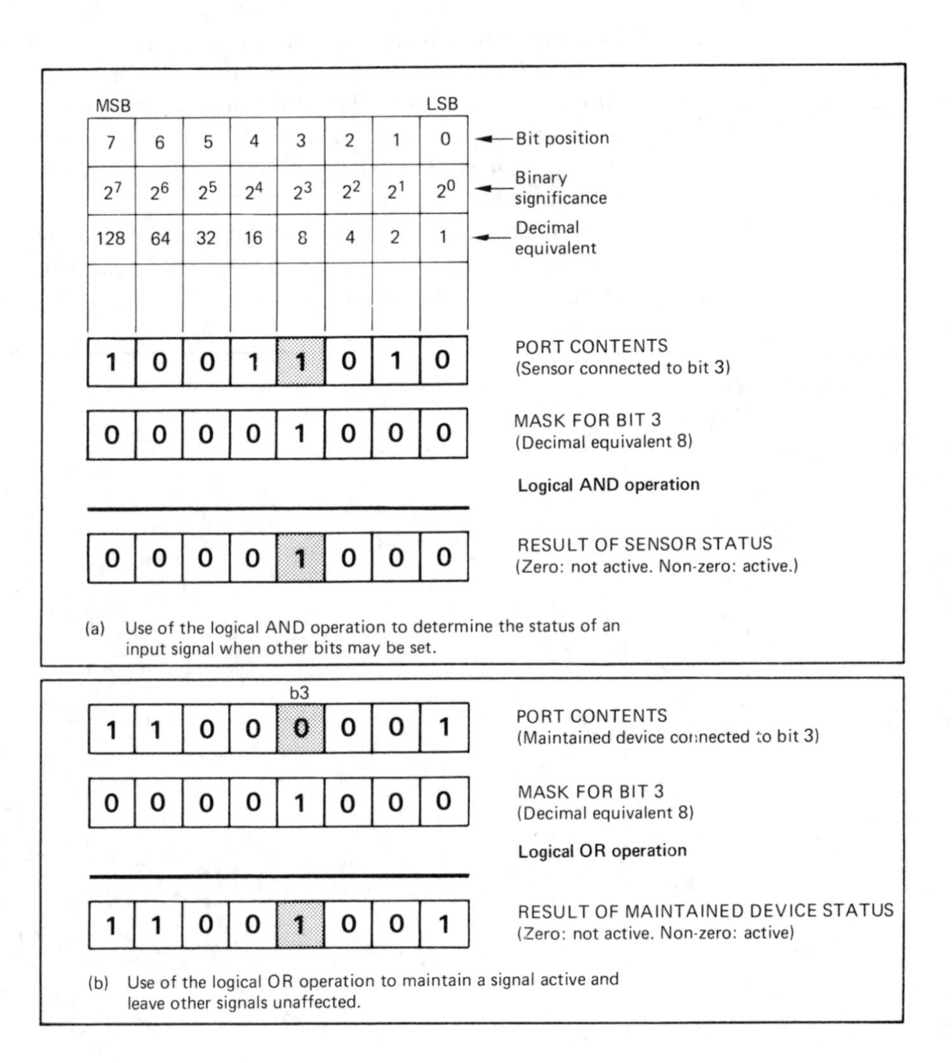

Logical OR operations can be implemented in a similar way. Logical OR operations are useful in applications where bits have to be intentionally set. For example, a software instruction equivalent to 'Port contents OR 8' will ensure that bit 4 is always set, irrespective of the port contents. (See Fig. 4/7*b*.) This may be useful when certain output signals have to be maintained, regardless of the actual port contents. The number, or **bit pattern**, with which the port contents are ANDed or ORed is known as a **mask**.

4.4 Data input

4.4/0 Punched tape

Punched tapes are predominantly used for data input in CNC machine tool applications. Since robots are a form of CNC technology, they may be used as a means of data input into a robot control system. Although punched tapes are not generally used in robot applications, it is instructive to study their characteristics and make-up.

Punched tapes for use in CNC applications are a standard 25 mm (1 inch) wide. They have a capacity for storing 10 characters per 25 mm. Thus, by measuring the length of the tape (in mm), dividing by 25 and multiplying by 10, it is possible to express the length of the control program by the number of characters it contains. Alternatively, the length of the control program may be expressed simply by quoting the length of the tape on which it is stored.

The right-hand edge of the tape is the **reference edge**. This is the edge adjacent to track 1. The offset of the feed holes (between tracks 3 and 4) helps to identify the reference edge. This offset ensures that tapes cannot be loaded into tape readers reverse way round. The direction of the tape can be identified in a number of ways. Many paper tapes have direction arrows printed on the upper face of the tape. In addition, when the paper tape is severed from its parent roll, the leading end will be pointed and the trailing end recessed. A typical punched tape is illustrated in Fig. 4/8.

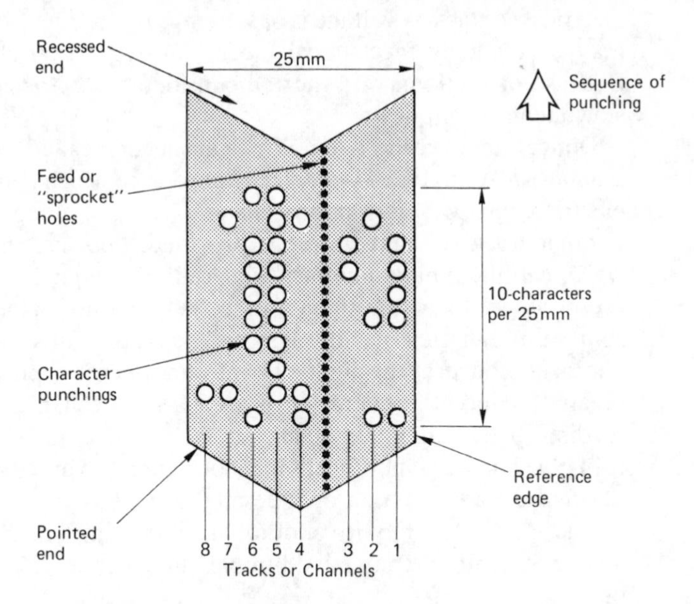

Fig. 4/8 A typical punched tape

Punched tapes may be read by mechanical or optical readers. The principle of operation of these devices is shown in Fig. 4/9*a* and *b*.

Mechanical tape readers employ eight spring-loaded wires or fingers corresponding to the tracks, or channels, running across the tape. The fingers bear against the tape as it passes through the reader. When a hole is

Fig. 4/9 Principle of operation of mechanical and optical punched tape readers

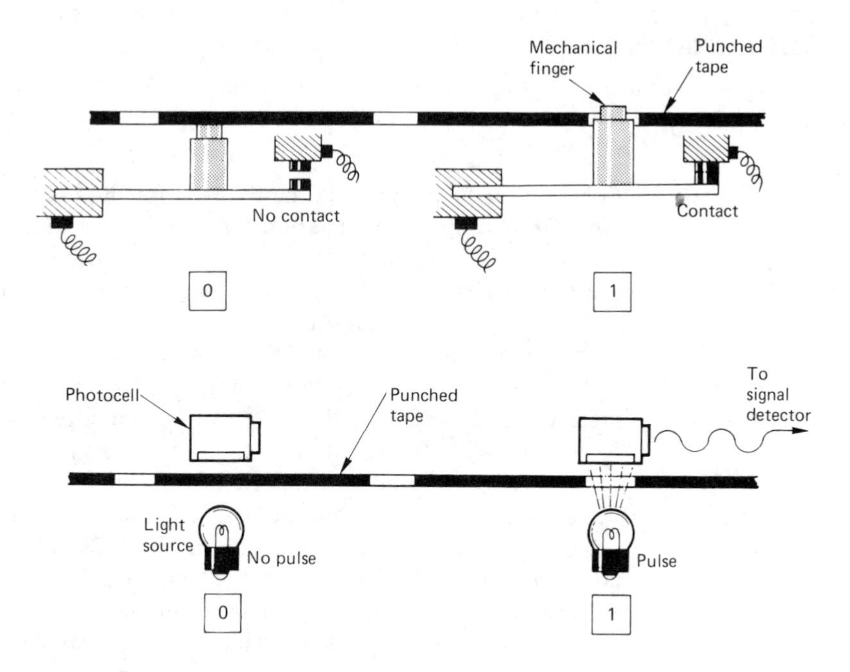

encountered, the finger momentarily passes through the hole and makes an electrical contact. A voltage is passed thus registering a binary 1. Punched tape may be read, by mechanical means, at around 50 characters per second. Because of frictional and inertial considerations it is considered a relatively slow means of input.

Optical tape readers employ transducers called *photo-electric cells* or *photocells*. A photocell is an electronic device which converts light energy into electrical energy. Eight photocells are arranged along the width of the tape, one per track. A light source is provided, opposite the bank of photocells, which can illuminate any particular cell. The tape is transported in between the photocells and the light source. When a hole is encountered, the light shines through the hole and the small beam of light strikes the photocell. The photocell converts the light dot into an electrical pulse which represents a binary 1. Since the operation is free from mechanical considerations, tape reading speeds can attain 2000 characters per second. For reliable operation, the light source and photocells must be kept clean and the punched tape should ideally be made from an opaque medium.

All tape readers require additional circuitry to synchronise read cycles with tape speed and, of course, additional punching devices are required to punch the media.

Although punched tape is often referred to as punched paper tape, a number of alternative materials are now being used which offer more desirable properties – for example, greater mechanical strength to lessen the likelihood of damage by tearing or worn sprocket feed holes, oil and water resistance, opaque and non-reflective properties, etc. Polyester or polyester/paper laminates are the common alternatives.

4.4/1 Manual data input (MDI)

Manual data input is the term given to data entry, into the robot control unit, via the console **keyboard** or **teach pendant**. Complete control programs may be entered at the machine by this method. A second, common application for MDI is the editing of control programs already resident in the controller's memory. This has the added advantage that, once edited, the program may be **downloaded** (saved) to the backing store or, if punched tape is being used, re-punched automatically by outputting to a tape punch.

4.4/2 Magnetic tape and disc

Magnetic tapes and discs are widely used in robot applications because of their proven design and usage with computers generally. They are reliable, robust, relatively cheap and can be built into original equipment as extremely neat and compact units.

They both apply the principle of storing data, in coded form, by means of *magnetised spots* on a magnetic medium. Both magnetic tape and magnetic discs are re-usable in that data may be erased and new data saved. Care must be exercised when handling, since dirt, grease or other foreign matter on the magnetic surface may cause data 'drop-out' resulting in inconsistent data transfer. Finally, data stored on both these media can be corrupted if they are brought into close proximity with magnets or stray magnetic fields.

Magnetic tape is a cheap and convenient way of storing large volumes of data in a comparatively small space. When the tape is housed in cassette form it is easy to handle, easy to store and is well protected. A typical cassette tape is 6 mm wide and can store around 100 characters per 25 mm. The data transfer rate depends on the speed at which the cassette drive operates. At 100 mm per second, for example, it can transfer up to 400 characters/sec. In computer terms this is relatively slow. Magnetic tape, like punched tape, is a **serial access** medium. That is, to isolate any piece of data on the tape it is necessary to read all the information before it. It is rather like music recorded on a musicassette. It is often quite difficult to pinpoint a particular piece of music accurately. This is because the tape passes the read/write heads in a continuous (serial) manner. Thus, if many control programs are recorded on a single cassette, some inconvenience may be incurred in locating the one required. Since the data is recorded magnetically, it is impossible for an operator to read the contents of the tape. In fact it is impossible, by inspection, to determine whether the tape actually contains information or not.

Magnetic tape recorders normally employ separate read and write heads. This enables the immediate reading of saved data, during write operations, which provides an early indication of data corruption or write errors. Data is usually recorded on 7 or 9 tracks across the tape width in **blocks** of 128 or 256 bytes along the length of the tape. Each block comprises a block identification header, followed by the data, followed by error check characters such as CRC (cyclic redundancy check) and LRC (longitudinal redundancy check). The format is illustrated in Fig. 4/10.

Magnetic discs, in contrast, are **random access devices**. That is, any single piece of data recorded on the disc can be accessed as easily, and as quickly, as any other. Common discs (because of considerations of cost) are the flexible, or **floppy discs**. These are circular discs and, like magnetic tape, consist of a plastic material coated with a layer of metal oxide which can be magnetised. The disc is enclosed in a square protective sleeve. The data is stored on concentric tracks which are arranged (electronically) on the surface of the disc. Standard sizes vary from 8 inch to 3 inch diameter with the 5.25 inch diameter floppy disc being most popular. Unlike magnetic tape, it is possible to use both sides of the disc for storage, if the correct hardware is available. Data is stored, in blocks, in concentric tracks formatted (electronically) on the disc surface. The tracks are divided radially (again electronically) into sectors. A read/write head moves across the surface of the

Fig. 4/10 Recording format for magnetic tape

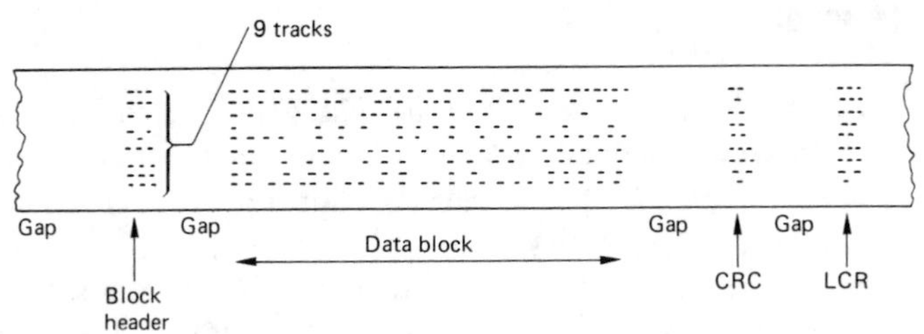

Fig. 4/11 Characteristics of magnetic discs

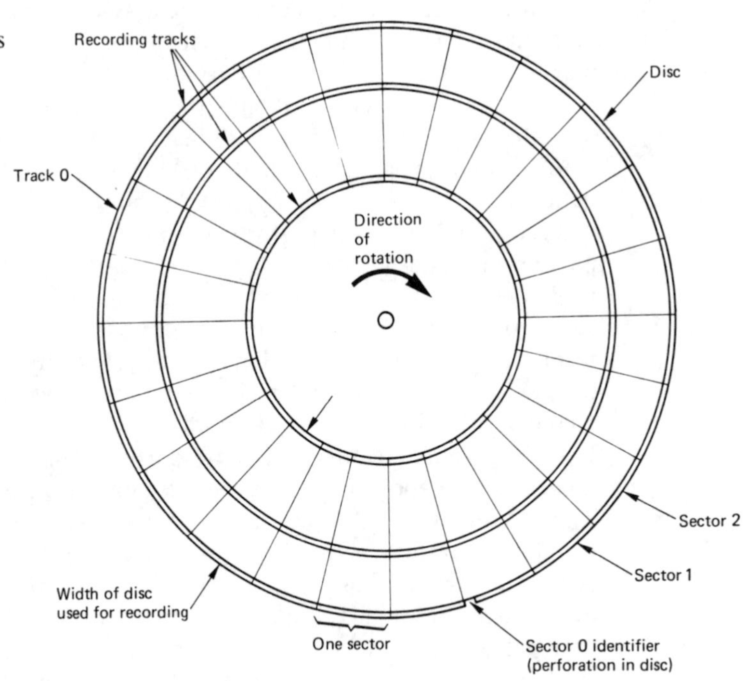

(a) Disc format

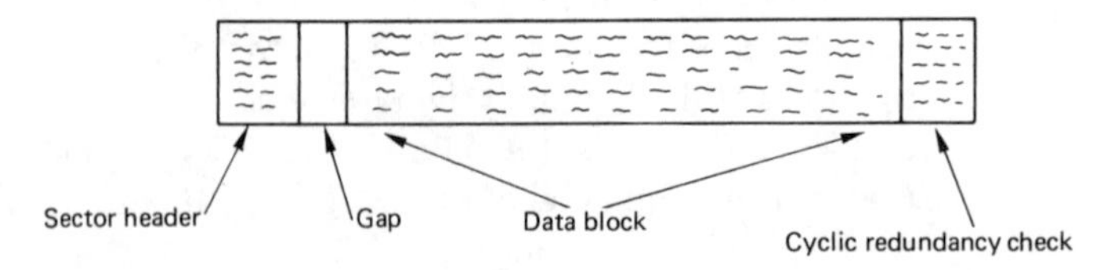

(b) Format of one sector

disc while it rotates (at about 300 rev/min). In this way it is possible to select just the piece of data required, rather like selecting a single track on an LP record. The capacity of floppy discs may range from 100K to 1200K characters. Data transfer rates are considerably faster than for magnetic tape. A typical transfer rate is 20K characters per second. The characteristics of disc storage are illustrated in Fig. 4/11*a*, *b* and *c*.

Typical backing stores and associated storage media are illustrated in Fig. 4/12.

Tape, disc and other peripheral equipment are relatively slow in operation, compared to the central processing unit. It is practice to feed such peripheral devices from a block of random access memory called a **buffer**. The CPU fills this buffer with the information it needs to transfer and can then resume its normal activities. The peripheral can then, simultaneously, carry out the transfer from the buffer. When the buffer is empty, the peripheral device signals the CPU to again fill the buffer. This is called **handshaking** and enables the CPU to continue with its processing unhampered by slow peripheral devices. Incoming data, from peripheral devices, also enters the control unit via the buffer. Buffers are quite separate areas of memory from the main internal memory of the control unit. A typical buffer size would be 512 bytes and a number of buffers may be active at any one time.

Fig. 4/11(c) (continued) Characteristics of magnetic discs

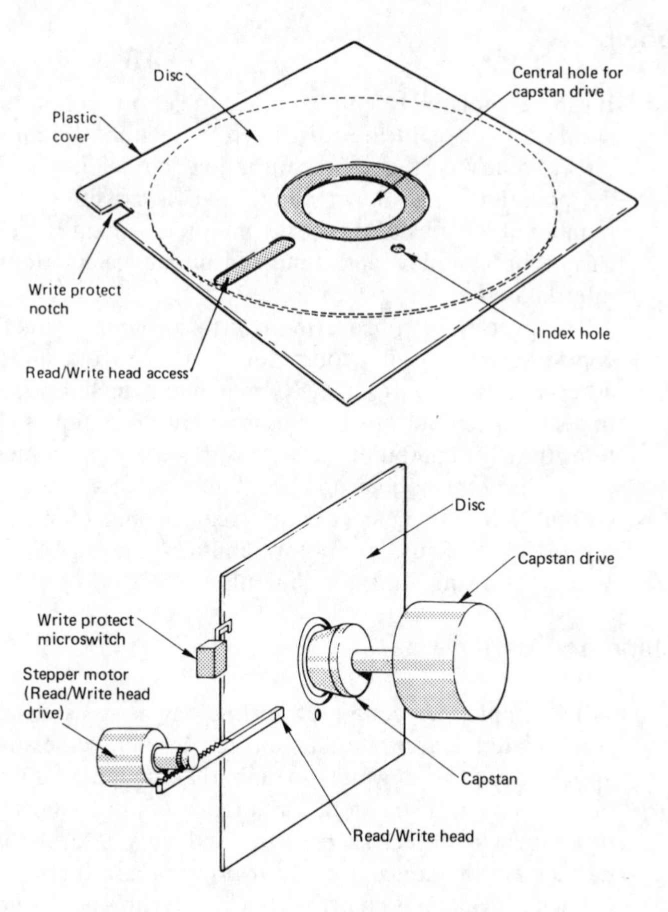

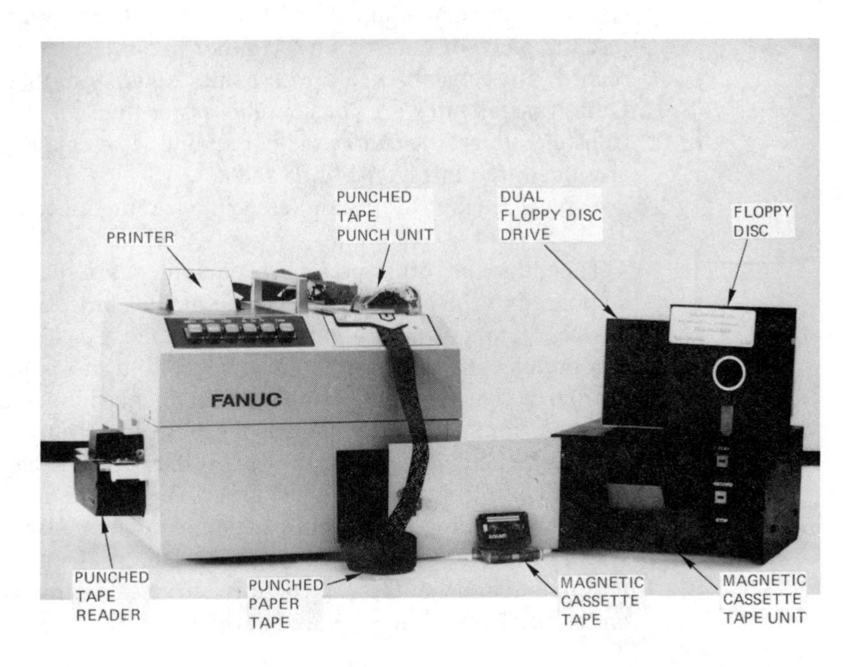

Fig. 4/12 Typical backing store and media devices [*courtesy: Aids Data Systems Ltd.*]

4.4/3 Host computer

It can be a relatively simple matter to get one computer to communicate with another. (This is further discussed in Sections 4.6 and 4.7.) Since transfer is direct, much of the time-consuming inefficiency and unreliability of slow peripherals is absent. In addition, it is possible to harness the vast computational abilities of the host computer (a mainframe or a minicomputer) to carry out complex and time-consuming calculations, such as calculating interpolated arm trajectories.

The process of transferring control programs into the memory of a robot controller from a host computer is called **direct numerical control** or DNC, a term borrowed from CNC machining techniques. This is the principle underlying remote off-line programming techniques. In many large installations the host computer has access to massive data files and may be linked to many different robots and machine tools. As one machine completes its current job, the host computer can arrange for the next program to be downloaded. Thus, a whole manufacturing installation may be under the control of a single master computer.

4.4/4 Error checking and parity

When employing computer technology it is essential that the data being manipulated is accurate and correct. In many cases the amount of data is so vast, and the transfer rates so high, that human operators cannot detect errors due to data transmission. A number of error-checking devices have been developed to detect such errors and duly inform the operator. Two such devices are the concept of a *parity check* and the use of *check digits*.

Parity checking is an error-checking technique designed principally for data stored and transmitted by punched tape. Although it is largely redundant in magnetic backing store devices, the principle behind its operation is interesting.

Many code systems utilise a 7-bit code. An eighth-bit is provided as a **parity bit**. The parity bit is a checking device used to determine whether a character has been coded, or transmitted, correctly. The parity bit is set to a 1 or a 0 so that the total number of 1s in the character (byte) is even (for **even parity**) or odd (for **odd parity**). Once again, it is convenient to refer to information coded in punched tape to illustrate the principle. For example, the ISO tape code specifies even parity. Reference back to Fig. 4/4 will confirm that the number of holes representing any character (along any row) will be even. If the code for a particular character results in an odd number of holes, then a hole is punched in the parity track, track 8, automatically by the tape punch. When data is read into the computerised control unit, a parity check is made on the incoming data. If a parity fault is detected, transmission will cease and the operator will be informed by some form of error alarm or error message. Common causes of parity faults and tape-read errors are shown in Fig. 4/13. Many control systems apply parity checks to incoming data however transmitted. Parity may be switched in, or out, or between odd or even to account for the various means of data input encountered. Since parity checking is obsolete when using magnetic storage media, many coding systems make use of the eighth-bit to decode extra characters.

Fig. 4/13 Common causes of punched tape read errors

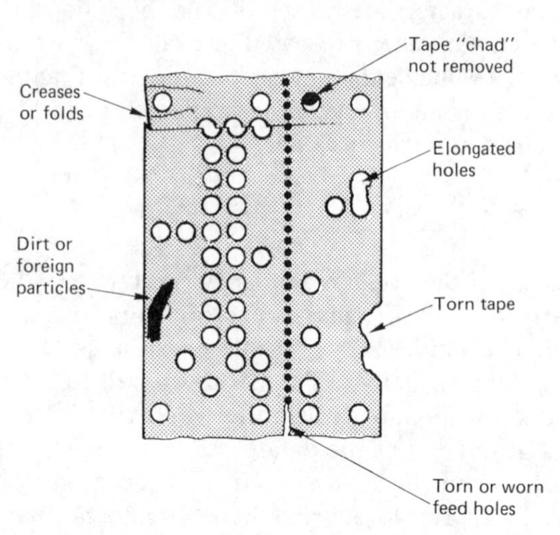

4.4/5 Check digits or checksums

Numeric or coded data items of particular importance often have **check digits** (somtimes called *checksums*) attached to them. The value of the check digit is determined from the characters that make up the data item, by using a simple formula. The system reading the data uses the same formula to compute the check digit. If the data item has been transmitted correctly, then the check digits will be the same. If not, the difference will be detected, transmission aborted, and the operator alerted by an error alarm or error message.

For example, to compute a check digit for the number 12345 we might proceed as follows:

1 Assign each digit a weighting factor:
   1 2 3 4 5 data item
   6 5 4 3 2 weighting factor
2 Multiply each digit by its weighting factor and add:
   $(1\times6) + (2\times5) + (3\times4) + (4\times3) + (5\times2) = 50$
3 Divide by 11 and note the remainder:
   50/11 = 4 remainder 6
4 Check digit is formed by subtracting remainder from 11:
   $11 - 6 = 5$
5 The check digit is placed as the least significant digit of the number:
   123455

When the number is checked, each digit is assigned the same weighting factor. The check digit is assigned a weighting factor of 1:

1 2 3 4 5 5 data item
6 5 4 3 2 1 weighting factor

Each digit is multiplied by its weighting factor and added:

$$6 + 10 + 12 + 12 + 10 + 5 = 55$$

If the resulting number is exactly divisible by 11 then the test is satisfied.

If alpha characters are present within the code, then the data items can be dealt with in a number of ways. For example, each alphabetic character can be assigned a value depending on its position within the alphabet (A=1, B=2, Z=26, etc.), or alternatively, each character can be processed on the basis of its decimal equivalent value.

4.4/6 Cyclic redundancy check

The success of the **cyclic redundancy check** (CRC) lies in probability theory and statistical analysis. It has a much higher probability of error detection than most other techniques. In principle, each block of data is considered as one very long binary number. The CRC is obtained by dividing this number by a second known number called the *generator*. The remainder of this division is stored at the end of the data block as the CRC value. The same check is performed when the data is read, and the newly calculated CRC value is compared with the stored value. Any discrepancy indicates either a saving or a transmission error.

4.4/7 Data transmission

Earlier in this chapter it was stated that computer memory can be visualised as rows of storage locations (likened to individual pigeon holes) that are capable of holding a single character or byte of information. Furthermore, each character is represented, in coded form, by eight binary digits or bits. These bits, in computer memory, are in fact electronic switches, which could be either switched on (set to binary 1) or switched off (set to binary 0). Thus, it can be inferred that each memory location is, in reality, a 'set' of 8 bits and that a bit being set to 1 is indicated by a small voltage present at that switch.

When data is being transmitted between a computer device and a peripheral (as with DNC, for example), it is done so by sensing (and saving) the voltage level of each individual bit of a particular memory location. These voltage levels are re-constituted, in the same order, and stored by the peripheral device. All data transfer is accomplished in this way.

Obviously the computer device will have to be connected to the peripheral in some way, normally by a cable. There are two ways of transferring the voltage, or bit pattern, which represents the data. The first, and perhaps the most obvious, way is to provide each bit with its own wire along which the voltage level can be sensed. This requires a cable comprising 8 wires, one for each bit. Such cable is supplied with the eight wires arranged side by side as a flat ribbon, and is known as **ribbon cable**. Because the voltage level of each bit appears on its wire at the same time as its neighbour, a byte of data can be transmitted in one go. This is known as **parallel transmission**.

The second way in which data transfer can be accomplished is by transmitting each item of data, a bit at a time, down a single cable. Although this will be slower, it has the distinct advantage that only one wire is required, or two if data transfer both ways is required. This may not seem to be that much of an advantage but it means that installations become cheaper and that data can be transferred over long distances, via existing telephone networks. Thus, local, national and even international exchanges of data can be accomplished with comparative ease. This system of transmitting data is known as **serial transmission**.

4.4/8 The RS232 interface

Although both methods of data transfer described above are widely used within computer systems, the serial transfer system has become standard in the industrial and commercial sectors. This is because of its compatibility with many major devices within the computer industry itself (large mini and mainframe computers, for example), and the obvious advantages associated with being able to transfer data down existing telephone networks.

If serial transfer is to be successful, there has to be some system whereby individual bits can be differentiated, and also where one byte of data stops and the next byte starts. Remember: serial data transmission appears, in practice, as a stream of voltage ONs and OFFs travelling along a single wire. An American (EIA) standard known as the RS232C standard for Serial Data Transmission has been adopted for this purpose. Within this standard there is facility for specifying parameters such as odd, even or no parity checking on transmitted data, the speed at which data is sent (the baud rate), and the format of the synchronising bits (start and stop bits) required to differentiate between different bytes of data.

It has to be arranged that the transfer speed of the transmitting computer is compatible with that of the receiving computer. This is termed the **baud rate** where baud means 'bits per second'. Transfer rates are usually selectable between 75 and 19 200 baud. As a rough guide, baud rate divided by 10 gives the transfer speed in characters per second. It must be stressed that the various settings on the receiving device must be completely compatible with those of the transmitting device for successful transfer to occur. For example, the transmit rate of the transmitting device must be set the same as the receive rate of the receiving device.

Normally, a collection of electronics can arrange that any output can be made serially. This collection of electronics is often available on a single plug-in circuit board known as an **RS232 interface**. The various parameters, mentioned earlier, may be changed either by software within a program, or by hardware selection by providing switches or links on the circuit board.

Nearly all industrial computerised control systems support an RS232C communication facility. It may be specified as standard equipment or may need to be specified as a fitted option. On some devices the RS232 interface may be termed a V24 connection.

Questions 4

1 State *five* principal advantages of employing computer-based systems.
2 Explain the difference between a 'general-purpose computer', a 'dedicated computer' and a 'microprocessor'.
3 Draw a block diagram of a typical computer system, stating the function of each sub-system.
4 State *three* typical input devices and *three* typical output devices that can be used in computer-based systems applied to engineering.
5 Define the following terms in the context of computer memory devices: 'RAM', 'ROM', 'byte', 'bit' and 'backing store'.
6 Explain why the binary system is so important in computer-based systems.
7 Convert the following decimal numbers into their binary equivalents: 1, 32, 157, 255, 129, 65 and 5.
8 Convert the following binary numbers into their decimal equivalents: 11001001, 11000001, 01000000, 00000000, 00000001 and 10101010.
9 Explain how alphabetic characters and other textual symbols can be represented within the binary system.
10 Discuss the relative advantages and disadvantages of storing robot sequence programs on punched tape and magnetic discs or tapes.
11 Explain the differences between ASCII and ISO binary coding systems.
12 Explain the concept of binary coded decimal in the representation of coded numerical data.
13 Explain the operation and use, in both hardware and software, of the logic operations AND and OR. Where could these find application in industrial robotic situations?
14 Briefly explain a common means of controlling input and output signals in computerised systems.
15 State *two* error-checking devices that can be used to detect input and data transmission errors and explain how they work.
16 State *four* means of inputting sequence program data into the control unit of an industrial robot.
17 State and explain *three* ways in which the length of a robot sequence program can be stated.
18 Explain the difference, outlining any advantages and disadvantages, between serial and parallel data transmission.
19 What is an RS232 interface and where would it be used?
20 What is meant by the term 'baud rate' and where would it be used?

End Effectors and Workhandling 5

5.1 End effector considerations

5.1/0 The need for end effectors

An **end effector** corresponds to the wrist and hand of an industrial robot. Since, in many robotic applications no actual holding or gripping takes place, the term 'hand' is inappropriate and the general term 'end effector' is used instead. Another more generalised term, *end-of-arm tooling*, may sometimes be used in cases where grippers are not employed. Some robot configurations may include one or more wrist motions within their basic movement capability. To avoid confusion and for convenience, wrist motions will be discussed as though they were part of an end effector. The comments are however transferrable should wrist motions be provided as part of the basic robot movements.

Once selected, an industrial robot (whatever its configuration) becomes the basic building block around which the performance of an intended task is based. Because it is a basic building block it is a general-purpose device that can be configured to perform one or more quite disparate tasks. The robot manufacturer cannot know to what tasks the robot will be assigned, and cannot predict the type of end effector required. Additionally, there is no universal end effector that will accommodate all likely applications. Consequently many robot manufacturers do not supply end effectors as standard equipment with their robots. They have to be specified as optional extras (most likely from a range of standard designs, sizes and applications), or be purpose-designed and custom-built. It must be inferred, therefore, that a robot does not become a functional system until it is equipped with a suitable end effector.

5.1/1 End effector mounting

Whilst perhaps not providing end effectors, the robot manufacturer does, however, have to ensure that there is some means of attaching end effectors to the end of the robot manipulator. The most common means of accomplishing this is for the end of the robot manipulator to be provided with a standard end effector mounting arrangement. This interface between the robot and the end effector is an important feature of the functional robot system. In addition to providing a physical mounting it may also provide information and service access to the functions of the end effector. The **end effector mounting** arrangement may have the following features:

1 It should provide quick connection and release.
2 It should not require an unduly accurate docking procedure when coupling up since automatic end effector changing may be required in mid-sequence.
3 It should maintain accuracy, strength and structural rigidity under both static and dynamic operating conditions.
4 It may provide connection of power service facilities such as pneumatics, hydraulics and/or electrics.
5 It may provide connections for control instructions and position and sensory feedback.
6 It should be tolerant of environmental conditions to which it may be exposed in the workplace such as oil, moisture, dirt, etc.
7 It should be resilient under occasional collisions.
8 It should be safe, for example in explosive environments.
9 It may incorporate either fail-safe or overload protection in the event of an unexpected or overload condition being encountered.
10 It should be wear-resistant for long life.

Each robot manufacturer is likely to have its own standard, but it is equally likely that this standard will be implemented across its entire range of robots. Thus, if an existing robot is replaced or supplemented by additional robots from the same range, the end effectors will be interchangeable and re-usable. This latter point is important since the end effector can represent a significant investment, sometimes as much as 25% of the original robot purchase price. Of course in other cases, the end effector may simply consist of a simple hook, ladle or some standard tool such as a spray gun, welding torch or hand grinder attached directly to the standard mounting plate. An example of an end effector mounting arrangement is shown in Fig. 5/1.

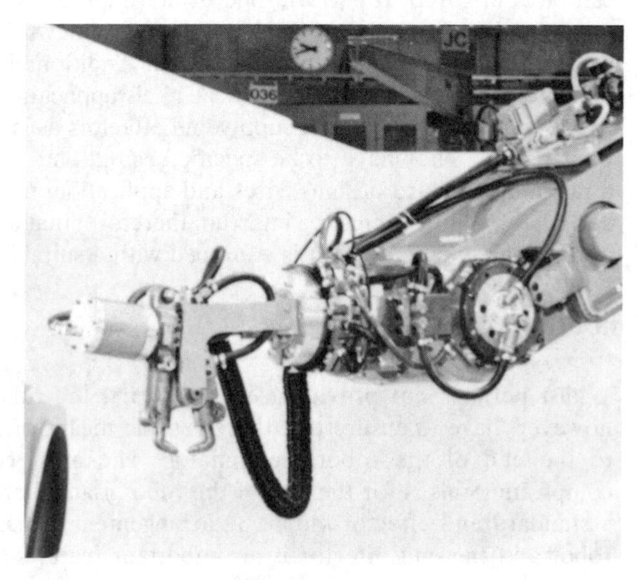

Fig. 5/1 An end effector mounting arrangement showing the connection of service supplies

5.1/2 Pitch, roll and yaw

It was stated in Chapter 2 that one function of the end effector might be to provide extra degrees of freedom to the basic robot configuration. It is common (although not essential) for end effectors to provide three degrees of movement and offer three degrees of freedom to supplement those already present within the basic robot configuration. In most cases the intention is to ensure that the final robot/end-effector combination can provide the full six degrees of freedom required to fully position a body in space. Or more correctly, to position and orientate its end effector at any position within the working envelope.

Generally, an end effector which offers these movements does so via rotary motion. In order to identify and differentiate between them, the movements are termed *pitch*, *roll* and *yaw*. Some of these motions may be provided as part of the main robot configuration itself. With reference to the human hand the movements can be illustrated as follows. If the hand is held with the fingers pointing forward in the horizontal plane and assuming a pivot about the wrist,

PITCH is angular rotation in the vertical plane. It can be demonstrated by pointing the fingers either up or down as if the hand were waving.
ROLL is angular movement about the axis of the arm. It can be demonstrated by turning the hand over either clockwise or counter clockwise in the manner of turning a door knob.
YAW is angular rotation in the horizontal plane. It can be demonstrated by allowing the hand to remain horizontal and pointing the fingers either to the left or the right.

These end effector wrist movements are normally provided by three separate revolute joints rather than by a multi-purpose joint such as a ball and socket type arrangement. This is illustrated in Fig. 5/2.

Fig. 5/2 End effector motions of pitch, roll and yaw

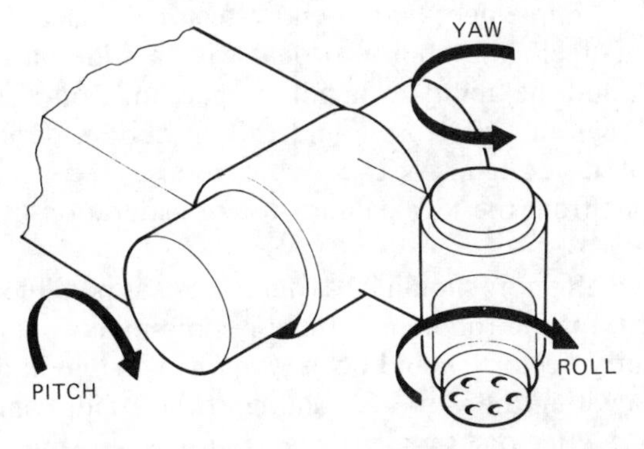

In cases where these motions are provided as part of the basic robot configuration they may be considered as axes in their own right. A common (although not standardised) means of identification of the movements is to designate them by a letter of the Greek alphabet as follows:

PITCH motion	β (beta) axis
ROLL motion	α (alpha) axis
YAW motion	γ (gamma) axis

The amount of rotary movement available within each joint is dependent on the physical design of the end effector. In some cases, movements may be restricted to less than 360° and in others they may be free to rotate through many revolutions. All the wrist movements require position measuring transducers to enable control of their position. In some cases sensing transducers will be provided within the end effector itself, for example force sensing to limit gripping forces. There will also need to be some convention for determining positive and negative movement of each motion.

The provision of a gripping action does *not* constitute a degree of freedom.

5.1/3 Considerations of end effector design

In making the human/robot comparison the end effector is analagous to the human hand. The design of end effectors, however, is best carried out strictly on the requirements of the task and not on considerations of human features and movements. Indeed in many cases, end effector designs can be far superior to the human hand. They can:

a) Utilise many means of handling cargo other than by mechanical gripping.
b) Be designed to offer superior gripping power and deal easily with heavy cargoes.
c) Be made tolerant of hot, corrosive, electrically active, sharp or rough cargoes.
d) Be made tolerant of hostile environments such as furnaces, chemical baths, gas or radioactive chambers, etc.
e) Be changed to suit different tasks and cargoes, even in mid-sequence.
f) Be designed to have superior manipulation and reach within confined spaces or during intricate tasks (e.g. paint spraying).

In contrast some very desirable attributes of the human hand are, at present, difficult to implement within end effector designs. These include the many senses of 'feel', the ability to deal with a wide variety of complex-shaped cargoes, and the ability to handle fragile and non-rigid materials.

The ideal end effector should exhibit certain desirable characteristics by virtue of its design. As ever, compromise and practical limitations force deviations from the ideal. Points for consideration include the following:

a) End effectors should be as *lightweight* as possible to increase the payload capacity of the robot/end-effector combination.
b) End effectors should be as *small* as possible in order to gain maximum access within the working volume of the robot configuration.
c) End effectors should be as *rigid* as possible to maintain accuracy, repeatability and prevent mis-handling of the cargo, whilst at the same time preventing damage due to over-gripping.
d) Heavy loads should be held as close as possible to the axes of movement of the end effector to reduce (force × distance) moments acting at the end of the manipulator, thus promoting structural instability.
e) End effectors should be *safe* both in their operation and in the event of overload or collision conditions.
f) End effectors should be available at reasonable cost.

As with other robot joints, end effector motions often incorporate some means of **braking**. The attainment of largely frictionless movements to reduce power requirements, and increase speed and smoothness of operation, also means that the joints are unable to sustain stationary positions under load conditions unless power is continually applied. To enable this holding power to be removed from the elements when the joints are stationary, braking systems can be incorporated within the design. Care has to be taken to ensure that the braking system is designed to fail safe. The usual approach is to apply power to hold the brakes off. Any loss of power thus results in the brakes being automatically applied. In order to reduce the power requirements further, it is common for the withdrawn brakes to be held off by means of, say, a solenoid-operated plunger or ratchet arrangement. Only a small amount of power is required to continuously supply the solenoid. Disc or drum type braking systems, similar to those used in motor cars, are common since they are efficient and compact and can easily be designed within revolute joints.

End effector design is engineering design at its most challenging. It embraces considerations of: mechanisms and forces; power application; materials selection; structural and dynamic behaviour; manufacture; costing; functional purpose and safety; within the constraints imposed by the task and workplace and under the influence of the points above.

Automatic end effector changes can be accomplished under program sequence control. The manoeuvre depends on alternative end effectors being available at a designated tool change station. Automatic quick-release operation and standard mounting arrangements facilitate uninterrupted operation requiring no human intervention. There are a number of reasons why such a facility may not be desirable. These include:

a) The time taken to change an end effector is non-productive.
b) Separate end effectors have to be designed and built and their whereabouts controlled.
c) Power services may have to be disconnected and reconnected.
d) Different end effectors may require different power services.
e) Continuous or prolonged end effector changing imposes harsh conditions on the end effector mounting arrangement.
f) It exposes extra potential problem areas such as damage, loss or errors concerning related end effector combinations.

An alternative solution to the problem, especially where the task is short in relation to the end effector change time, is to utilise **multigrippers**. These designs incorporate a number of grippers, tools or sensors within a single end effector. They are provided on a rotatable turret which can be indexed to bring any individual element into position or, alternatively, they provide duplicate gripping elements. Care must be exercised in cases where the elements in dormant positions may interfere with the task being performed by the active element. Also if the multigripper is large, long withdrawals may have to be made in order to perform indexing without fear of collision. A common example is that of a robot loading and unloading, say, a turning centre. The component is likely to be of different size and shape after the machining operation than when it was loaded. If a single gripper cannot be designed to accommodate the difference then a two-stage multigripper would be ideally suited. A second example is that of an external gripper and an internal gripper provided within the same end effector unit. Examples of multigripper type end effectors are illustrated in Fig. 5/3.

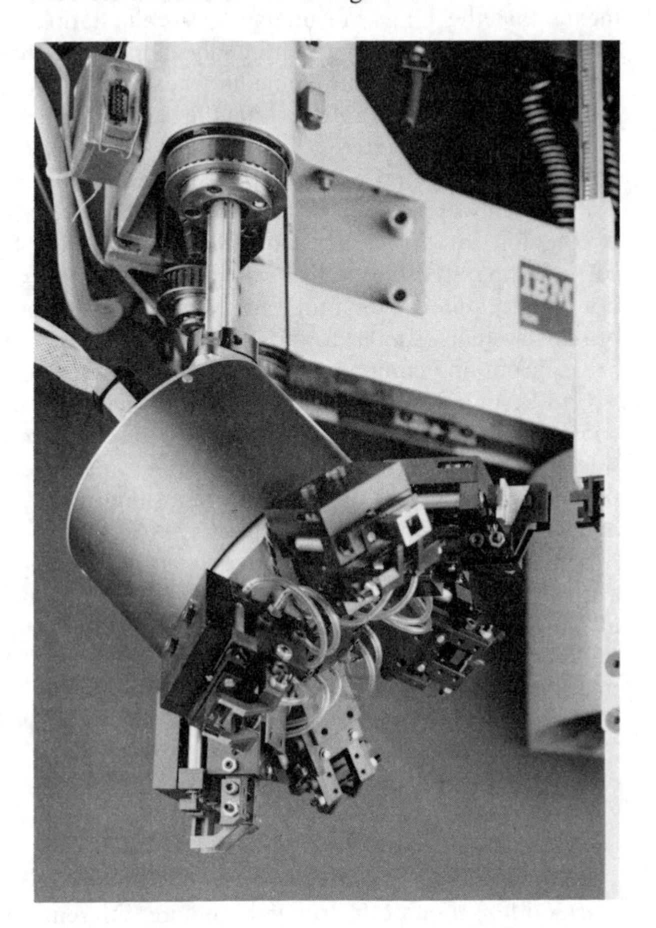

Fig. 5/3 Multigripper type end effector fitted to a SCARA robot configuration [*courtesy: IBM*]

End effectors may be constructed from a number of materials depending on the nature of the task, the environment within which they are required to operate, and the form of supply of the material. For example, steel offers strength and rigidity and is available in a number of easily obtainable forms; aluminium and plastic offer lightness, anti-magnetic properties, and resistance to oxidation in easily obtainable structurally efficient sections; and ceramics offer resistance to heat, chemicals, acids and corrosion. Both plastic and ceramic materials are good insulators. Materials such as polyurethane are popular coatings for gripper jaws since they have a high coefficient of friction for non-slip grip, and they offer slight compliance under compression. This elasticity enables the coating to mould itself to the shape of a component, and it will retain its shape under repeated applications of such compression and expansion. A combination of materials make up most end effectors and many bought-out parts are often incorporated. Examples include bearings, sensors, leadscrews, gears, cylinders, air motors and so on. In hostile, dirty or corrosive environments it may be necessary to protect or shield the end effector and the robot by other materials and devices.

5.2 Types of end effector

5.2/0 Classification of end effectors

End effectors may be broadly classified as grippers, tools or sensors depending on their design and intended purpose.

Grippers represent by far the majority of end effector types. They are characterised by being able to grasp, manipulate, transport and position a cargo. Although they do not necessarily have to be mechanical in operation, like for example the operation of the human hand, many grippers do incorporate tactile (touch) sensor elements to feed back information regarding the operation or status of the gripper, or to discriminate between different components. For example, sensors can be incorporated to monitor if, and how tightly, the gripper is holding a cargo, or to determine the size and/or shape of the cargo. Gripper design, whether mechanical or otherwise, must consider:

a) Size of the cargo.
b) Shape of the cargo.
c) Weight of the cargo.
d) Orientation/presentation of the gripper to the cargo.
e) Strength of the gripper in relation to applied forces.
f) Strength of the cargo in relation to the applied forces.
g) Uniformity of gripping and localised pressure spots.
h) Likelihood of marking the surface of the cargo.
i) Positioning of the cargo such as self-centring.
j) Safety in operation and use.
k) Cost of manufacture.

Types of gripper are discussed in the following sections of this chapter.

Tools include a broad range of devices. They may be powered tools such as grinders, drills and nut runners; or they may be applicator devices such as spray guns, nozzles and welding torches; or just simple utensils such as hooks, cups, ladles, scoops and piercing spikes. Many tools require a power source and most applicators require some means of supplying the consumable product. The bulk supply of a dispensed product may have to be carried by the robot itself, or it may be provided remotely by some umbilical arrangement from a fixed storage location.

Sensors take many forms. They may be stand-alone sensors such as proximity (presence) sensors and those used to check for leaks in car windscreen assemblies; or sensors incorporated within other end effector types, for example force sensing in mechanical grippers or joint following when using, say, a welding torch. Sensors and sensory feedback are discussed in Chapter 6.

5.2/1 Mechanical grippers

Mechanical grippers are the most common form of gripping device. The majority of designs are modelled on clamping mechanisms. The efficiency of mechanical grippers depends on the applied force, the coefficient of friction between the jaws and the cargo, and the nature of contact, i.e. point, line or surface. The **gripping force** can be calculated thus:

Gripping force = Applied force × Coefficient of friction

Note that the gripping force is independent of the area of contact. The stability and rigidity of the held component, however, may be improved by increasing the area of contact by altering the mode of application of the force. This is illustrated in Fig. 5/4 which shows the effect of gripping a cylindrical object using point, line and surface contacts. For these reasons it is common for the jaws either to be shaped to suit the form of the cargo (e.g. vee-shaped for gripping circular components) and/or to be surfaced with a high friction material (to increase grip) or compliant material (to 'mould' to the cargo shape).

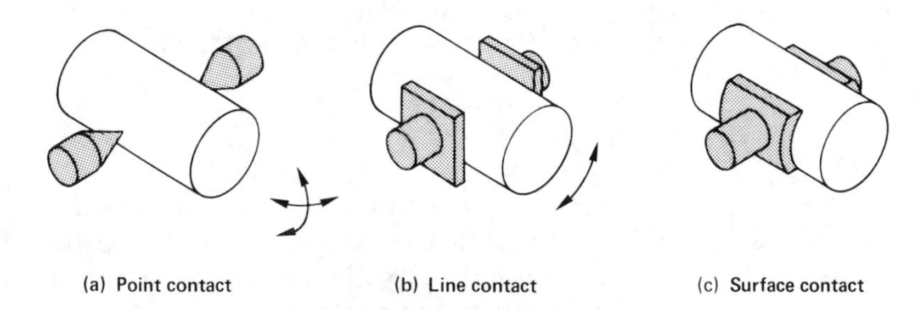

(a) Point contact (b) Line contact (c) Surface contact

Fig. 5/4 Contact conditions for mechanical grippers

Mechanical grippers normally require the movement of one or more elements (usually some form of jaw) under the control of some (power assisted) mechanism to achieve a gripping action on the cargo. The provision of tapered rubber cones within the jaws can enable components to be grasped by means of existing holes. This may be more convenient when the cargo has a non-uniform or unpredictable external or internal shape, but contains holes of the same or varying diameter in a known position. An advantage of mechanical grippers is that the same end effector can be utilised in different applications by simply changing gripper jaws.

There are a number of ways of classifying mechanical grippers.

One classification identifies **coarse grippers** or **precise grippers** depending on the accuracy and sophistication of their construction and their intended purpose. Coarse grippers are found in applications where components have to be held, transported and manipulated often within large spans of movement and with relatively little accuracy. Examples include the loading and unloading of furnaces, stacking and palletising applications, and the dipping or quenching of components during coating or heat treatment processes. Their design and operation is likely to be simple and unsophisticated. Precise grippers ensure the accurate holding, positioning and orientation of the cargo being manipulated. They are deployed in the loading and unloading of machine tools and other manufacturing processes, pick and place applications, and some assembly operations. By definition their design and construction are accurate and efficient.

A second classification distinguishes between **hard grippers** and **soft grippers**. Hard grippers (the most common type) normally incorporate jaws of a fixed size and shape. The jaws offer either internal or external gripping, are roughly contoured to suit the shape of the cargo, and incorporate some means of self-centring. They are however rigid and offer no 'give' (technically termed *compliance*) to any deviations in the shape of the cargo. Any such deviations usually result in an inferior grip being applied and possible movement of the cargo during its manipulation. Conversely, soft grippers are designed to take up the shape of the cargo. These work by surrounding or enveloping the shape of the cargo, or being inserted *inside* a vessel, before pressure is applied to expand the gripper to achieve the gripping action. Soft grippers are useful in circumstances where:

a) A number of different shaped components need transporting, without requiring time-consuming end effector changes.
b) The gripper needs to be tolerant of changes in size or shape of a single component.
c) It is desirable to equalise the gripping pressure on fragile components.
d) Marking of the component surfaces must be reduced to a minimum, or eliminated.

Fig. 5/5 An example of a soft gripper

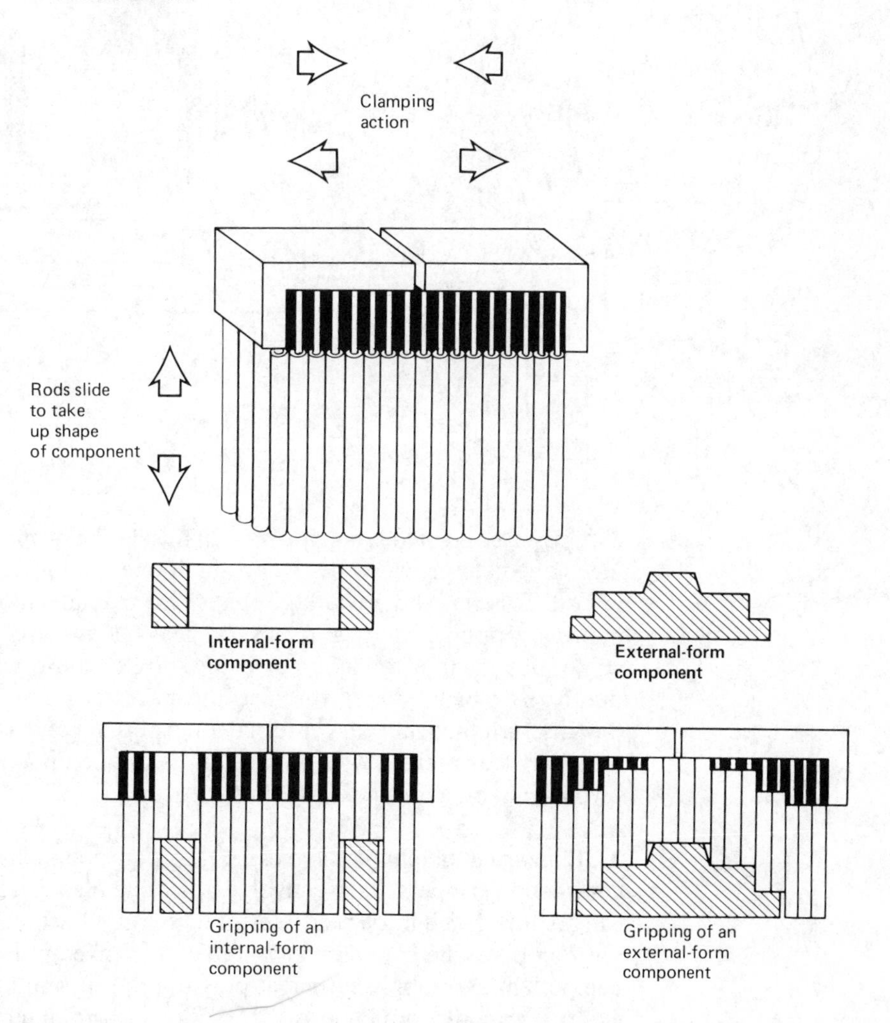

Whilst soft grippers have been demonstrated successfully they are not sufficiently developed to be in widespread industrial use. An example of a soft gripper is shown in Fig. 5/5.

A third classification relates principally to the design and operation characteristics of the gripper. Three examples are offered below.

1 Parallel or translational jaw grippers In this design, usually two jaws accomplish gripping via a linear or parallel motion relative to each other. The design may move both jaws towards and away from each other or there may be one fixed and one moving jaw. Gripping action may be external (when the jaws close together on the outside of a component), or internal (when the jaws move away from each other within some internal feature of the component), or both. Examples of parallel motion grippers are illustrated in Fig. 5/6.

Fig. 5/6 Principle of parallel motion mechanical grippers

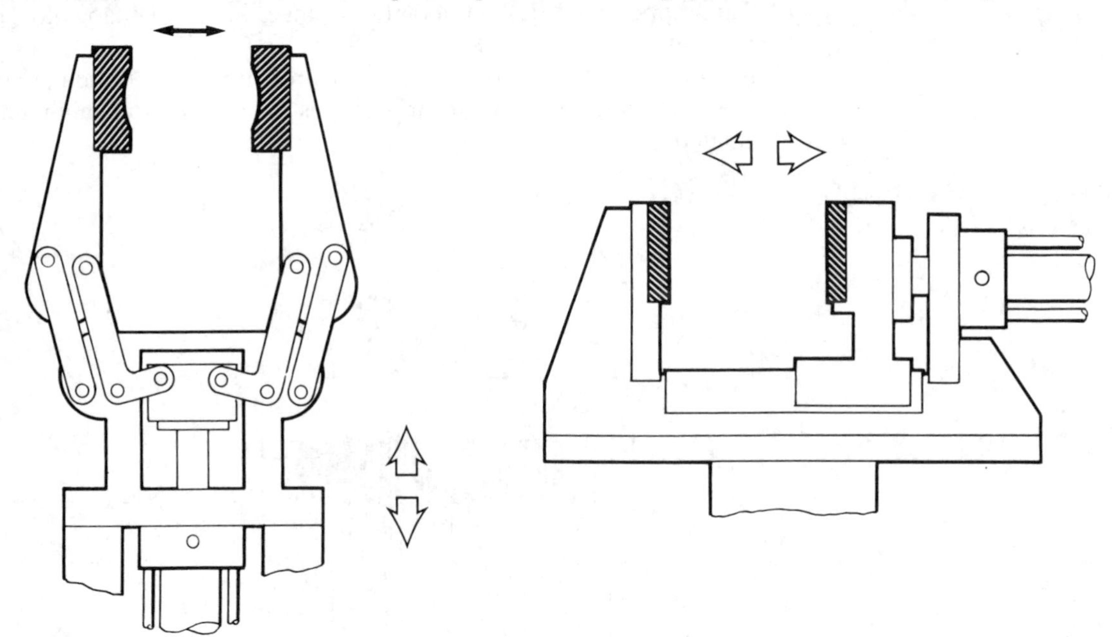

2 Scissor or pivoted jaw grippers In this design, gripping action results from jaw movement provided by the action of one or more links operating through pivots. This has the advantage that force magnification can be achieved to improve gripping characteristics. If heavy loads are lifted vertically using scissor designs they are inherently safe since the downward force of the cargo, due to its weight, tends to close the jaws together. As with translational designs, gripping may be internal or external, and return springs may be incorporated where power application is in one direction only. Examples of pivoted jaw type grippers are illustrated in Fig. 5/7.

3 Expansion grippers Typical of soft grippers are the expansion and contraction grippers. These involve inflatable bladders which, when deflated, can be introduced into holes or cavities provided within the component cargo. Once in place the bladder can be inflated to take up the internal shape of the component and supply sufficient pressure to grip and lift the component. For external gripping, similar loose bags, filled with granular materials such as sand

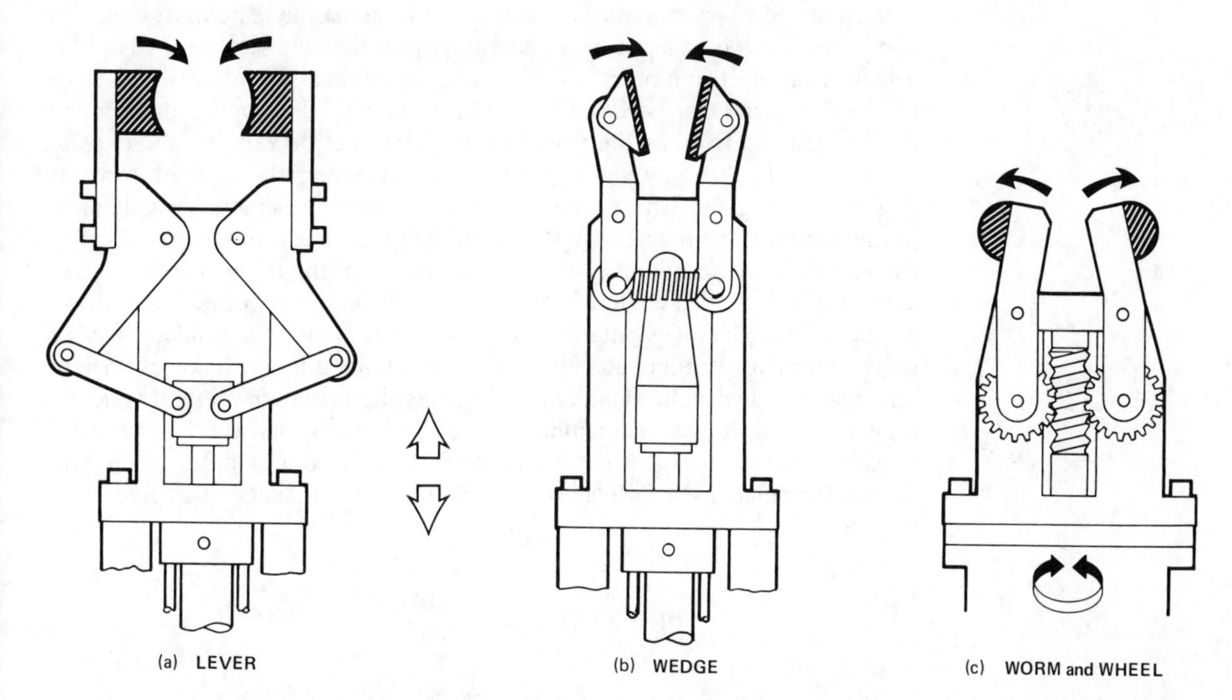

Fig. 5/7 Principle of pivoted jaw mechanical grippers

or magnetic powder, can be draped over the component cargo. An applied vacuum or electro-magnetic field gives the granular material enough rigidity to hold the form of the component, and enough strength to hold and/or lift it. These grippers are especially suited to those applications where fragile and delicate components (such as glass containers, sand cores, etc.) need to be handled without fear of breakage, and where the shape of the cargo is complex or unpredictable. Purely mechanical expansion grippers can be configured using elastic materials and causing them to expand to effect gripping. Expansion grippers do have slight disadvantages. It is difficult to incorporate sensing techniques to predict whether a component is present or not. Since the position of the gripping action is not precisely defined, it is also difficult to predict the exact position of the held component relative to the end of the robot arm. Examples of expansion-type grippers are illustrated in Fig. 5/8.

Fig. 5/8 Expansion grippers

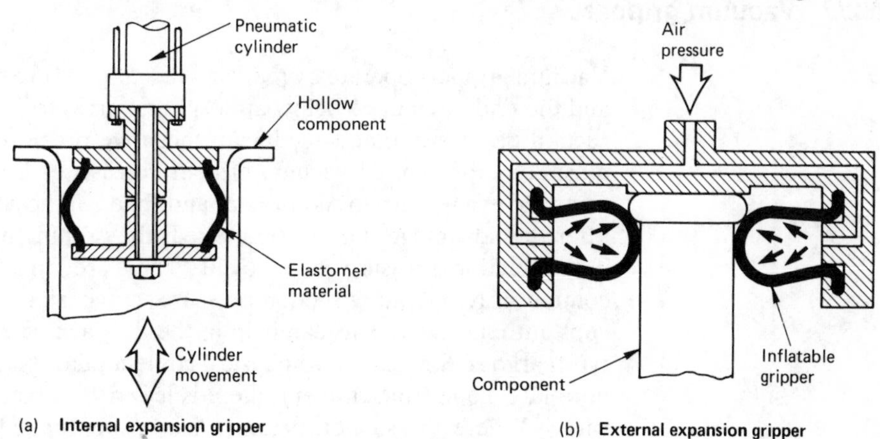

Mechanical grippers require some means of **actuation**. The most common applications involve the use of pneumatics, hydraulics and screw thread drives, driven from electric motors. (The relative advantages and disadvantages of these systems are discussed in Chapter 3, Section 3.4.) Gripping force can be determined and fixed by the design of the gripper or be variable and depend on sensory feedback. One example of determining the gripper force is illustrated in Fig. 6/2, where rollers provide the gripping element. Any slippage of the component (indicating insufficient gripping force) causes the rollers to rotate due to the frictional contact. The rotation of the rollers can be sensed and used to inform the end effector that more force is required. In the case of mechanical linkages, mathematical techniques using considerations of forces, moments, torque, etc. are used to design the gripping linkages. Where pneumatic or hydraulic cylinders are used as the actuation method, calculations involving the cross-sectional area of the piston, and the line pressure applied to the cylinder, can determine the force exerted. Examples of common design formulae used, which will be familiar to students of engineering, include:

$$\text{PRESSURE} = \frac{\text{FORCE}}{\text{AREA}}$$

$$\text{MOMENT} = \text{FORCE} \times \text{PERPENDICULAR DISTANCE}$$

$$\text{CLOCKWISE MOMENTS} = \text{ANTI-CLOCKWISE MOMENTS}$$

$$\text{UPWARD FORCES} = \text{DOWNWARD FORCES}$$

$$\text{TORQUE} = \text{FORCE} \times \text{RADIUS}$$

$$\text{FORCE} = \text{MASS} \times \text{ACCELERATION}$$

5.2/2 Vacuum grippers

Vacuum grippers operate by reducing the air pressure between the component and the end effector to below atmospheric pressure. More correctly they are suction devices in that a significant 'negative pressure' is often applied. The most common type of vacuum gripper consists of a number of **suction cups** arranged in a pattern to suit the size and shape of the component. Suction cups offer the advantage that, once applied, they retain a grip on the cargo even if the suction pressure is removed. They are, in effect, fail-safe. With a continuously operating suction pressure, cargoes are attracted to the suction cups automatically. This can help in the de-stacking of some laminar (sheet) type cargoes. Since no hard physical contact takes place (the suction cups are normally made from rubber), there is less risk of marking and damaging the cargo. Where the suction pressure has to be applied intermittently, a time penalty is incurred whilst air is removed and the suction pressure builds up.

Both techniques require some means of detaching the cargo from the suction cups to enable their release.

The suction pressure can be achieved either by a discrete vacuum pump or by continuously flowing air causing a venturi effect. The latter approach lends itself to the use (in most factories) of readily available compressed air supplies. An air-tight seal between the suction cup and the cargo is not necessarily required so long as the suction pressure is sustained and large enough. Advantages of using vacuum type grippers include the following:

a) They are often simpler (hence cheaper and more reliable) than mechanical grippers.
b) The exact size and shape of the cargo is relatively unimportant.
c) No precise positioning of the end effector is required.
d) There is no danger of crushing or applying high localised 'pinch' forces to the cargo.
e) It is only necessary to have access to one side of the cargo.
f) They are safe in explosive environments.

They have the limitations that the cargo must offer a smooth, clean and unperforated surface large enough to accommodate the diameter of the suction cup(s); there must be provision for a pump or external air supply; and the level of noise may be irritatingly loud and unceasing. It is common for suction cups to be mounted on springs. The reasons for this are threefold:

a) Used in conjunction with limit switches it enables a 'search until found' strategy to be employed where the position of the target cargo cannot be accurately predicted, or where the height of a stack of components continually reduces.
b) It reduces the harshness of contact that could arise with a heavy or fast-moving arm.
c) It allows a suction pressure to build up whilst the arm is still decelerating onto the target, thus minimising the idle time in waiting for the suction pressure to build up. The time required for a vacuum to build up may be important in applications where robots are engaged in servicing continuously operating processes.

Lifting capacity of vacuum grippers is dependent on the size of the suction cups and the efficiency of the vacuum produced. It can be approximated using the following simple formula:

$$\text{Lifting capacity (kg)} = \frac{0.785 \times \text{Percentage vacuum} \times \text{Diameter of suction cup}}{100}$$

In cases where multiple cups are supplied from the same source, the above result must be multiplied by the number of cups used. The percentage vacuum will vary depending on, for example, whether venturi or pumps are employed and the extent of sealing present at the suction cups. Care must be exercised when using several suction cups, supplied from the same source, to lift a number of separate components. The absence of one component will mean that none of the components is lifted because of the inability to support the vacuum.

An example of an end effector employing vacuum grippers is illustrated in Fig. 5/9.

Fig. 5/9 End effector employing vacuum grippers

5.2/3 Magnetic grippers

Magnetic grippers are usually based on the principle of electromagnetism, although permanent magnets can be used for smaller cargoes. An electromagnetic field is a circular magnetic field generated around an electrical conductor when an electric current flows through that conductor. The magnetic field has a direction associated with it depending on the direction of the applied current. It follows that reversing the direction of the applied current reverses the direction of the electromagnetic field. This principle can be used to repel cargoes in order to help release them from the magnetic gripper. The strength and efficiency of the electromagnetic field, and hence the lifting capacity of magnetic grippers, can be increased by providing a large number of turns on the conductor arranged in the form of a coil, and even further by placing an iron core within the centre of the coil.

Magnetic grippers share many of the same advantages as vacuum grippers with the qualification that the cargoes must be ferromagnetic in nature.

a) They are simpler (hence cheaper and more reliable) than both vacuum and mechanical grippers.
b) The exact size and shape of the cargo is relatively unimportant, although flat components are preferred.

c) Magnetic grippers can be designed to suit the form of unusually shaped cargoes, which improves lifting efficiency.
d) Precise positioning of the end effector to the cargo is usually not required.
e) There is no danger of crushing or applying high localised 'pinch' forces to the cargo.
f) It is only necessary to have access to one side of the cargo.
g) The gripping action acts instantaneously on contact with the cargo and there is no time delay.
h) They are safe in explosive environments.
i) The cargo may be perforated although this reduces handling capacity.

Magnetic grippers exhibit residual magnetism which means that some way of **detaching the cargo** from the gripper is often required. When electromagnets are employed a short burst of reverse current normally releases the grip. Permanent magnet grippers require some mechanical means of releasing the cargo such as ejector pins or stripper plates. An alternative means of removing a cargo from a permanent magnet is to arrange for the magnetic poles to be shifted to redirect the magnetic field through the gripper and not the component. This principle is illustrated in Fig. 5/10 and is firmly established in the magnetic chucks commonly used on surface grinding machines. Highly ferrous cargoes could become slightly magnetic after being subjected to the magnetism of the grippers. Permanent magnets tend to become demagnetised at temperatures greater than 100° C, which may limit their use in some applications. Since magnetic force decreases with distance, care may have to be exercised with sheet materials which may tend to flex under their own weight and increase the distance between a sheet and the gripper.

Depending on the environment, some difficulty may also be encountered in keeping magnetic grippers clean and free from unwanted magnetic foreign particles such as swarf, iron filings, small screws, springs, etc. Also the working environment must not be sensitive to the magnetic fields exhibited by the gripper itself. Large flat components may slide along the surface of the magnet unless locations are provided to guide the cargo and resist its motion.

5.2/4 Difficult cargoes and operations

It is impossible to categorise the suitability of all cargoes within a range of standard end effector or gripper types. End effector and gripper design is largely an individual undertaking specific to the task, the cargo and the workplace environment. It may be necessary to combine or completely reject the concepts of the more common types of gripper described above, in formulating an acceptable solution to a given problem.

A good illustration of difficult cargoes concerns the handling of *cloths, fabrics and thin laminates*. These are examples of cargoes that are non-rigid; come in different thicknesses, textures and weights; are delicate, difficult to grip, and prone to damage and marking. They often come in bundles or stacks of multiple layers which are difficult to separate and tend to stick together even under lifting by vacuum (or in the case of fabrics, by piercing with needles or other puncturing tools). They are often difficult to maintain smooth and flat, and have different surface textures. Solutions to such problems include the

Fig. 5/10 Principle of permanent magnet and electro-magnetic grippers

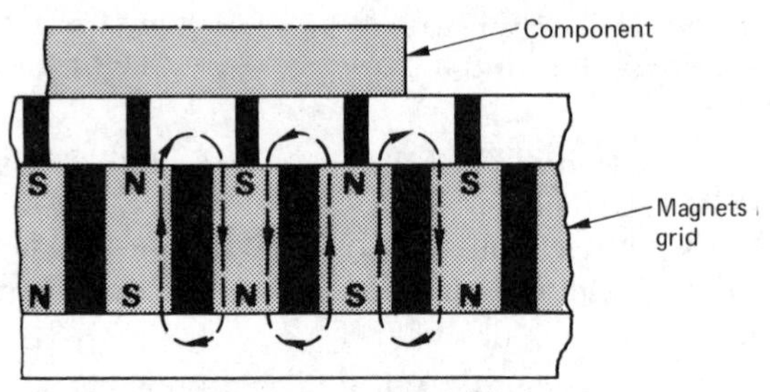

(i) RELEASE: magnetic flux bypassed by pole pieces

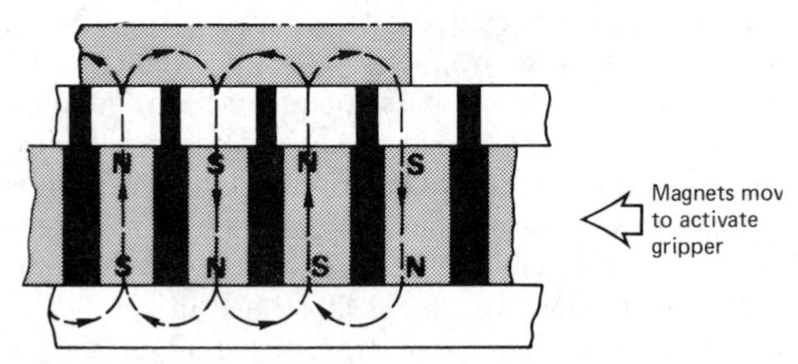

(ii) GRIP: magnetic flux passes through component

(a) **Principle of a permanent magnet gripper**

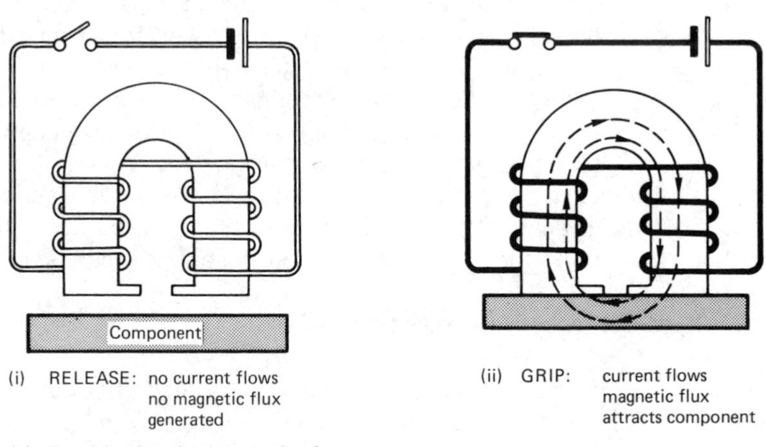

(i) RELEASE: no current flows no magnetic flux generated

(ii) GRIP: current flows magnetic flux attracts component

(b) **Principle of an electromagnetic gripper**

use of adhesive discs or tapes (but hot environments may cause the adhesive to dry up), electrostatic attraction, and the use of blow lifters. A jet of air blown across the surface of a layer of the cloth or laminate will cause the top layer to lift. The resulting vibrations set up by the air flow tend to stop subsequent layers from sticking. With semi-rigid laminates, air jets can be directed between or underneath the various layers causing them to rise by aerodynamic lift.

Assembly tasks are difficult operations for robots to perform. Indeed assembly tasks are often difficult for human operators. This is acknowledged in the fact that male elements of an assembly are often provided with chamfers or tapered leads, and holes with similar countersunk chamfers. The human assembly operation relies heavily on the sense of 'feel' and a number of attempts may be required, by the operator, to achieve successful assembly. Research into assembly tasks carried out by robots has led to the development of special end effector devices with built-in compliance in order to emulate the dexterity of the human operator during assembly operations. They are known as **remote centre compliance** (RCC) **devices**. The RCC device attaches to the end of the end effector in between the end of the robot arm and the gripper. They can therefore be employed in tasks other than assembly operations. They are designed to provide a mechanical solution to correct mis-alignment rather than relying on expensive and complex sensors and high-accuracy mechanisms.

When assembly tasks are analysed, two types of error can be identified. These are *lateral positioning error* and *angular mis-alignment*. Lateral positioning error requires lateral movement to correct it, without any angular movement. Angular mis-alignment, which is a common cause of jamming and wedging, requires angular correction without lateral movement. These two conditions are illustrated in Fig. 5/11 showing a shaft being assembled into a hole.

Fig. 5/11 Assembly error conditions of lateral and angular mis-alignment

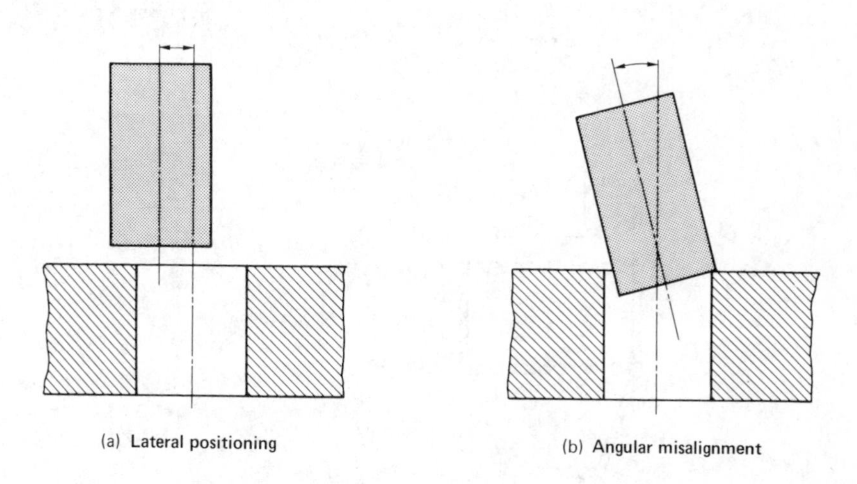

The RCC device allows each condition to be corrected individually and independently of each other, permitting successful assembly to occur. The principles of operation of the unpowered movements are illustrated in Fig. 5/12. This enables robots with relatively low accuracy specifications to perform assembly tasks which would otherwise require expensive sensors and sensing software.

In similar circumstances the stated accuracy of the robot can be increased by providing the end effector with its own fine positioning mechanism that can be fine-tuned in response to sensor feedback. The robot traverses as near to the commanded position as allowed by its accuracy and repeatability specification, and then hands over control to the end-effector/sensor combination. The end effector, under its own motive power and under the control of on-board positional sensing devices, will then achieve the final target position.

Fig. 5/12 Principle of the remote centre compliance end effector

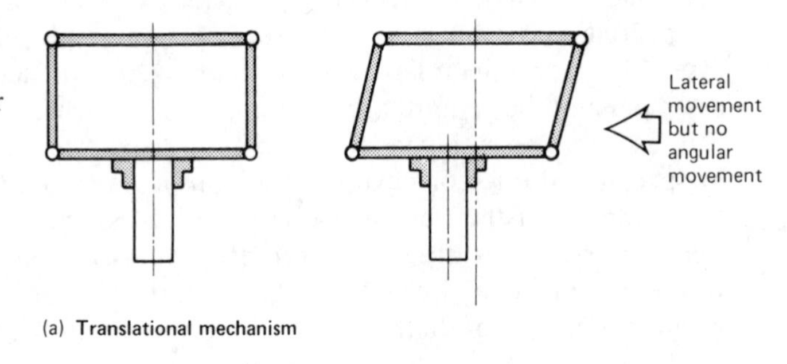

(a) Translational mechanism

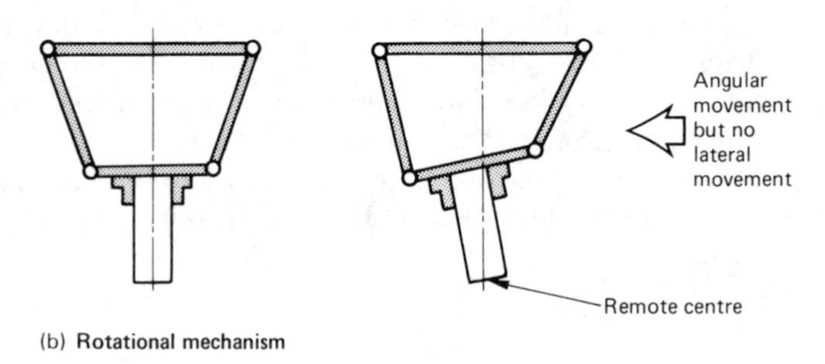

(b) Rotational mechanism

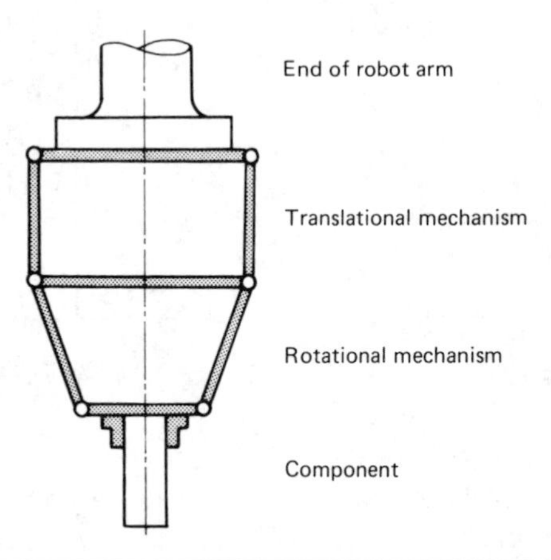

5.3 Workhandling

5.3/0 Workplace layout

It may be the flexibility of the robot, or the very limitations of its reach and movements, that will significantly shape the layout of the workplace environment within which the robot works. First and foremost it has to be able to reach those elements it is intended to service. A number of alternatives may be explored. They include:

a) Siting the elements requiring service in close proximity to the robot (or vice versa).
b) Providing the robot with locomotion to reach the elements requiring service (or vice versa).
c) Employing intermediate handling equipment (conveyor, AGV, etc.) to connect the robot with the elements requiring service.

These alternatives also have to be considered within the wider context of safety in the workplace. Common solutions to date have concentrated largely on fencing off robot workplaces and designating them human 'no-go' areas. The implications of safety, and the fact that both options *b* and *c* incur significant extra costs, extra considerations of control, and additional complexity, tends to favour option *a* as the first to be explored.

Fig. 5/13 Re-siting of machine tools for robot layout

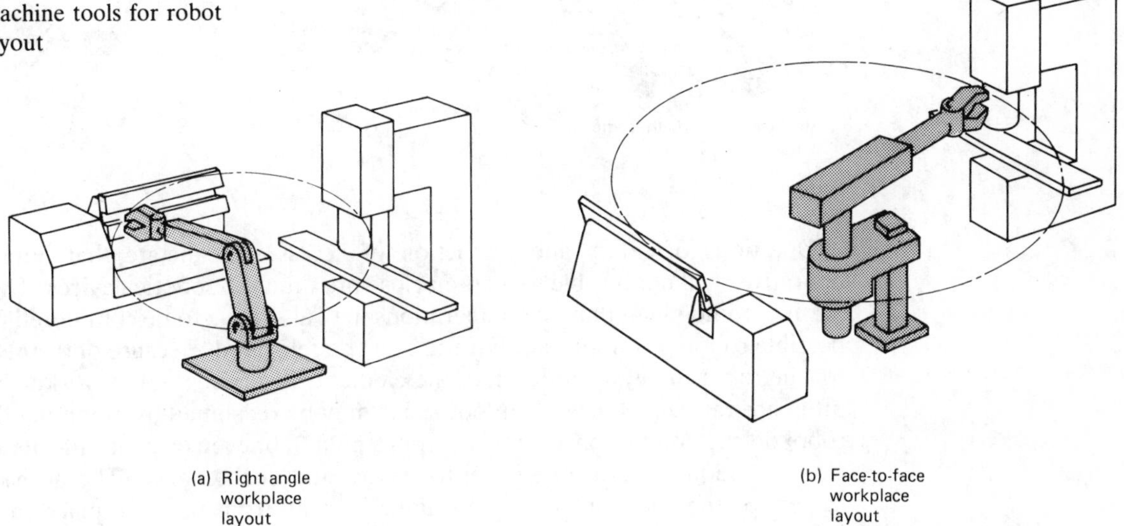

(a) Right angle workplace layout

(b) Face-to-face workplace layout

Within the ambit of option *a* it is common for machine tool layouts, for example, to be already established. Rather than attempting to select a robot to service the existing layout, it is very worthwhile to consider the re-siting of the machine tools in question. Consider the existing layout of two CNC machine tools shown in Fig. 5/13*a*. The specification of a robot to transfer components from one machine to the other would require a robot of the dimensions indicated. By re-siting the machine tools and mounting the robot on a rigid pedestal, as shown in Fig. 5/13*b*, a smaller, simpler and less-expensive robot may be specified. Whilst this simple example deliberately ignores factors of payload capacity, control system, accuracy, repeatability and so on, it does serve to illustrate the prudence of considering options.

The task of machine loading and unloading may dictate that the robot is spending much of its time idle, in waiting for long cycle times to be completed. This suggests that another task could be performed by the robot during these periods, or that more machines could be brought within the service of the robot. Either implementation will have implications for workplace layout. Some examples are shown in Fig. 5/14.

Fig. 5/14 Robot workplace layouts

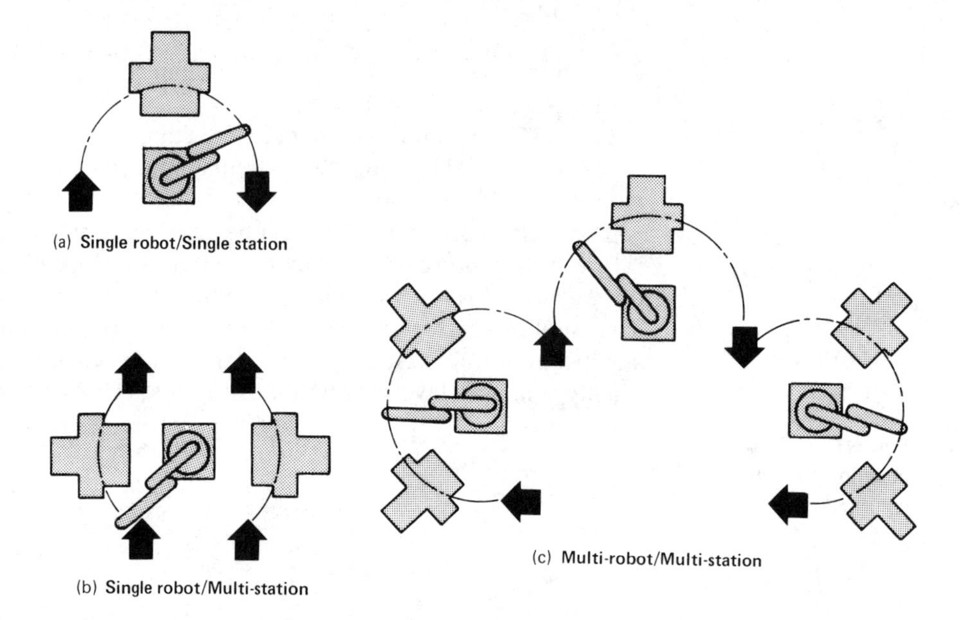

Questions of safety and restriction of access may dictate that human operatives will not be able to work in close proximity to robotic devices. This, in turn, could mean that allied operations may also have to be considered for possible robotisation, or that they too must be relocated. Because of the need for the robot to be within reach of the elements it services, robot workplaces often appear very congested. Floor space may be reclaimed by mounting the robot off the ground (on a gantry or from a pillar), or even on a machine itself. Consideration must also be given to service and maintenance. The decision to service the robot in situ, or to move it away from the workplace, may influence the final workplace layout and space utilisation.

The acceptance that a major malfunction of the robot could occur, either during its normal operation or during maintenance or programming periods, should also influence the design of the workplace. Attempts should be made to minimise both damage to human life and destructive damage of plant and equipment in the event of such a malfunction. Fundamental steps may include the following:

a) Minimising the number of potential trapping points.
b) Safety cages in the event of cargoes being dropped or thrown.
c) Intrusion monitoring systems to restrict human access.
d) Fail-safe mechanical devices, such as springs, shear pins, steel support cables, etc., to reduce the effects of heavy masses and rapid movements.
e) Inexpensive and easily replaced ancillary equipment, such as storage and work surfaces, etc.
f) Adequate training and familiarisation of personnel likely to have access to the workplace.
g) The establishment of safe and rigorous procedures with tell-tale indications of any departures from them.

The effects of any change in work patterns and any possible or anticipated future requirements should be such that expansion of the automated workplace entails minimum disruption and re-organisation.

5.3/1 Task analysis

Most robotic applications fall into three distinct broad categories. Those tasks that can be termed pick-and-place tasks; those tasks that require a gripper on the end of the robot arm; and those tasks that require some sort of tool or appliance attached to the end of the robot arm. In researching the introduction of robots into the workplace however, a more rigorous and comprehensive approach is applied.

The impetus to consider the introduction of robots into the workplace arises from a number of commercial influences. They include:

1 The inception of new products, processes or factories.
2 As part of a management-led strategy.
3 Production engineering factors such as quality, productivity, repeatability, etc.
4 Influences of external competition, towards greater flexibility, cost effectiveness, etc.
5 Customer pressure.
6 Prestige and/or the nature (usually high-tech) of the business.

Whatever the initial reasons, the selection of a particular robot configuration, a particular end effector and a particular workplace layout should be preceded by a comprehensive analysis of the task, or tasks, to be performed. Such an analysis is often termed a **feasibility study** and is required for the following reasons:

a) It will clarify the aims and objectives of the exercise.
b) It may indicate that robotisation is inappropriate to attain the aims and objectives being considered.
c) It may point to other areas in which improvements would increase the benefits to be gained, e.g. workplace re-organisation, re-design of components, automation of allied operations, etc.
d) It will force the implications of the exercise to be addressed in terms of forward planning and the integration and coordination with other aspects of the organisation, etc.
e) It will highlight the true costs of robotisation including factors of ancillary equipment, training, safety, programming, maintenance, production engineering support, human acceptance and re-deployment, etc., and determine the proposed payback period.
f) It will determine a specification for the robot and end effector.
g) It will propose an organised timetable of events for the planned implementation of the scheme.

A procedure for the implementation of a feasibility study into the introduction of robots should include the following:

1 Establish provisional aims and objectives based on the reasons for considering the scheme.
2 Assess the present situation and requirements together with future planning and forecasting information.
3 Formulate a list of alternatives for the achievement of **1** and **2**.
4 Acquire complete relevant information concerning each alternative.
5 Carry out an analysis and evaluation of each alternative including a detailed cost analysis.
6 Review the original aims and objectives in the light of **5** and decide on the most beneficial option.
7 Formulate a specification for the equipment and a sequence and timetable for the proposed implementation of the chosen scheme.
8 Arrange on-site demonstrations of the preferred robot system using standard production components to assess its suitability.
9 Review the objectives and the proposed implementation and return to step **1** if appropriate.
10 After implementation of the system carry out an evaluation to assess the success or otherwise of the installation based on the original aims and objectives.

Some or all components of the above procedure could well entail the use of external consultants and detailed discussion with a variety of suppliers. Where possible, scaled-down feasibility trials involving physical set-ups or research and development programmes should be carried out before final decisions are taken. Turnkey systems may be considered which may themselves incorporate the feasibility study element.

At the workplace level, task analysis will be required for the selection or design of a particular end effector. Such analysis may start from consideration of the tasks likely to be required of a robot/end-effector combination. Ten examples are

1.	UNSCRAMBLE	6.	ASSEMBLE
2.	SEPARATE	7.	TRACE
3.	GRASP	8.	LOCATE
4.	MANIPULATE	9.	POSITION
5.	ORIENTATE	10.	TRANSPORT

This type of analysis approach is useful in that it may also be used to help formulate a logical sequence of events to perform a given robot task. Additional considerations of coordination and interlocking with other elements of the workplace, and activating robot-held tools or applicators, must also be included. Indeed some high-level off-line programming languages include statements similar to the above functions in order to simplify the task of programming the robot.

5.3/2 Component analysis and design for automation

One of the useful aspects of considering robotisation and the design and operation of end effector devices is that it encourages objective analysis of the work task. In cases where transportation and manipulation of components are involved this analysis can extend to the design and potential re-design of the components themselves. This useful exercise is often termed **value analysis**. Whilst it may be carried out during a robotic feasibility study primarily concerned with component handling, it can often result in significant cost and component reductions, especially where assemblies are involved. Value analysis is predominantly carried out retrospectively on components that already exist. Taken to its logical conclusion, and where robot handling is consciously planned, design for *component handling* should be considered at the component design stage. This suggests that designs requiring as few degrees of freedom as possible to position and orientate the component are preferred.

As experience is gained with robotics it becomes more and more clear that robots and end effectors (unless equipped with complex sensory devices) are not suited to identifying and dealing with *randomly oriented components*. It follows that some means of identification and orientation of the components will be necessary, often by additional equipment provided at the workplace, before the robot can perform effectively. This equipment represents a hidden oncost and its complexity and expense to the installation will depend on how easy it is to identify and manipulate the components involved. Simple design points may enable this task to be performed quickly, easily and at minimum cost.

Two examples of *rationalisation of components* for an assembly operation are illustrated in Fig. 5/15 *a* and *b*. The assembly shown in *a* reduces the number of components to be handled from 4 to 2. It also illustrates that the relatively simple-to-implement task of rivetting may be more appropriate for robotisation than the difficult task of parts assembly. The solution shown in *b* eliminates the need for feeding and locating the washers prior to the screwing operation.

Fig. 5/15 Rationalisation of component design for automation

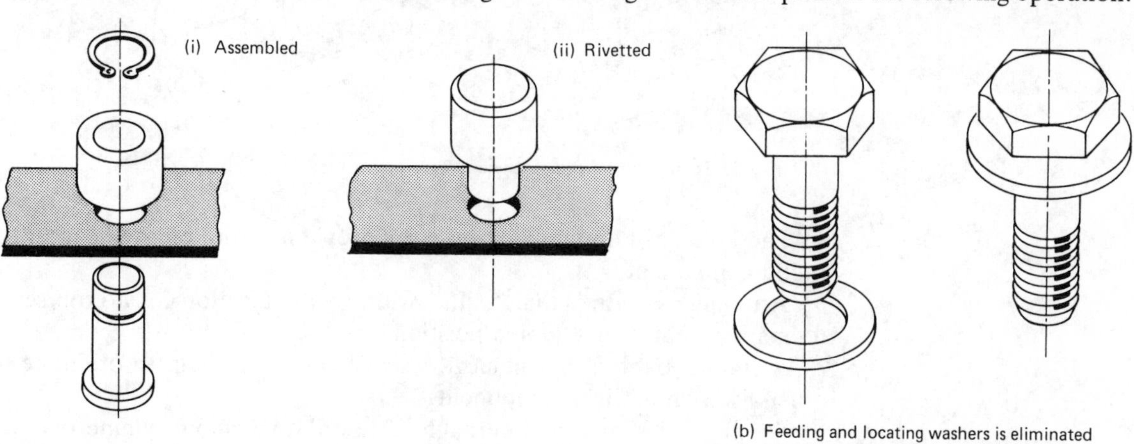

(a) Rivetting reduces the number of components from 4 to 2

(b) Feeding and locating washers is eliminated by using combined screws and washers

Symmetrical components reduce the need for exact identification and orientation. In cases where components are not symmetrical it may be prudent to provide a salient locating, orienting or handling feature. It may even be cost effective for such features to be sacrificial in that they can be removed and discarded when they have served their intended purpose. Four examples are illustrated in Fig. 5/16 *a*, *b*, *c* and *d*. The examples illustrate the following points:

Fig. 5/16 Component re-design for automation

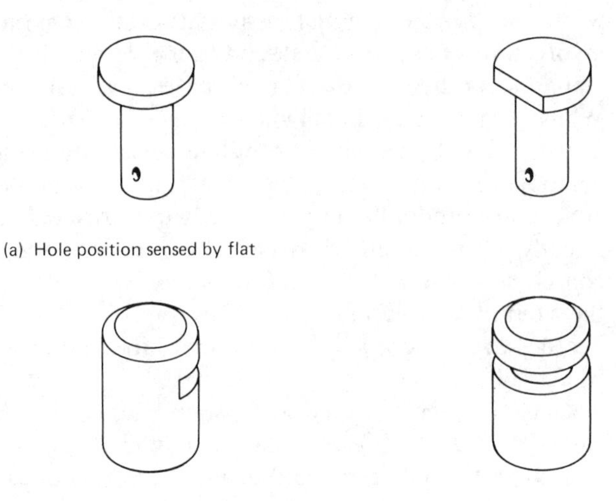

(a) Hole position sensed by flat

(b) Grub screw located in any position by annulus

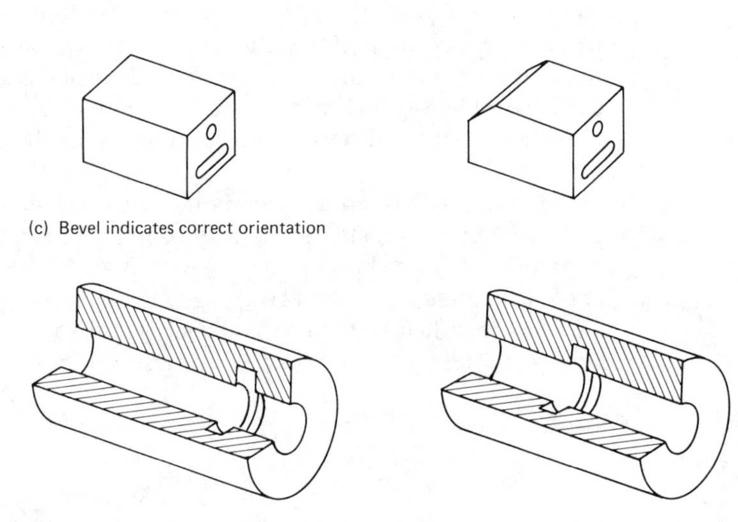

(c) Bevel indicates correct orientation

(d) Internal groove positioned to ensure symmetry

a) The hole position may be sensed by providing a flat on the head of the round component.
b) An annulus (rather than a slot) will provide location for a grub screw no matter what its assembled position.
c) A simple external mechanical design feature will help to orientate an otherwise symmetrical component.
d) Simple re-design of an internal feature will bring about symmetry of the turned component.

Of course the safe and functional purpose of the component must be retained throughout any re-design that takes place.

5.3/3 Parts presentation

The previous section has highlighted one of the major hidden costs of robotisation, that of parts presentation. **Parts presentation** is the term given to the production engineering function of ensuring that the parts being processed are presented to the robot and its end effector in a known position, with a known orientation, at the correct time and at a suitable rate. The parts may be discrete components or assemblies that have to be manipulated by the robot, or components that have to be fixtured such that the robot can perform tasks on and around them using tools other than grippers. It makes little sense to employ a robot that relies on human labour to feed it each component, and efforts are directed to automating the process. Effective parts presentation requires that three conditions be met:

1 That the parts are designed for ease of manufacture.
2 That the parts are manufactured within their specified tolerances.
3 That an adequate (continuous) supply of parts is available.

Parts presentation commonly involves two stages: the **orientation** of the components and the **feeding** of the components. In continuously fed operations, involving small components, both stages can be performed within a combined feeding device. A number of solutions to parts presentation, manipulation and feeding are readily available off the shelf and others may have to rely on special-purpose designs which are custom built. The choice of method depends on a number of factors:

a) The size, shape and weight of the parts.
b) The form of supply of the parts.
c) The desired output of the presented parts.
d) The mode of delivery of the presented parts.
e) The fragility of the parts.
f) The cost and cost effectiveness of the proposed solution.
g) The reliability of the proposed solution.

Important factors in considering combined methods of both feeding and orienting components are the volume required and the mode of delivery. For example, the parts can be fed continuously, continuously but at a paced rate, or on request from the robot or some other device in the workplace. Common methods include the following:

a)	*Feeders*	Reciprocating
		Vibratory (bowl or in-line)
		Conveyor
		Elevator
		Rotary
		Chute
		Tube
b)	*Magazines*	Dispenser
		Bandolier
		Strip
c)	*Pallets*	Stacks
		Boxes
		Trays
d)	*Hoppers*	

Fig. 5/17 Mechanical component orientation devices for continuously fed components

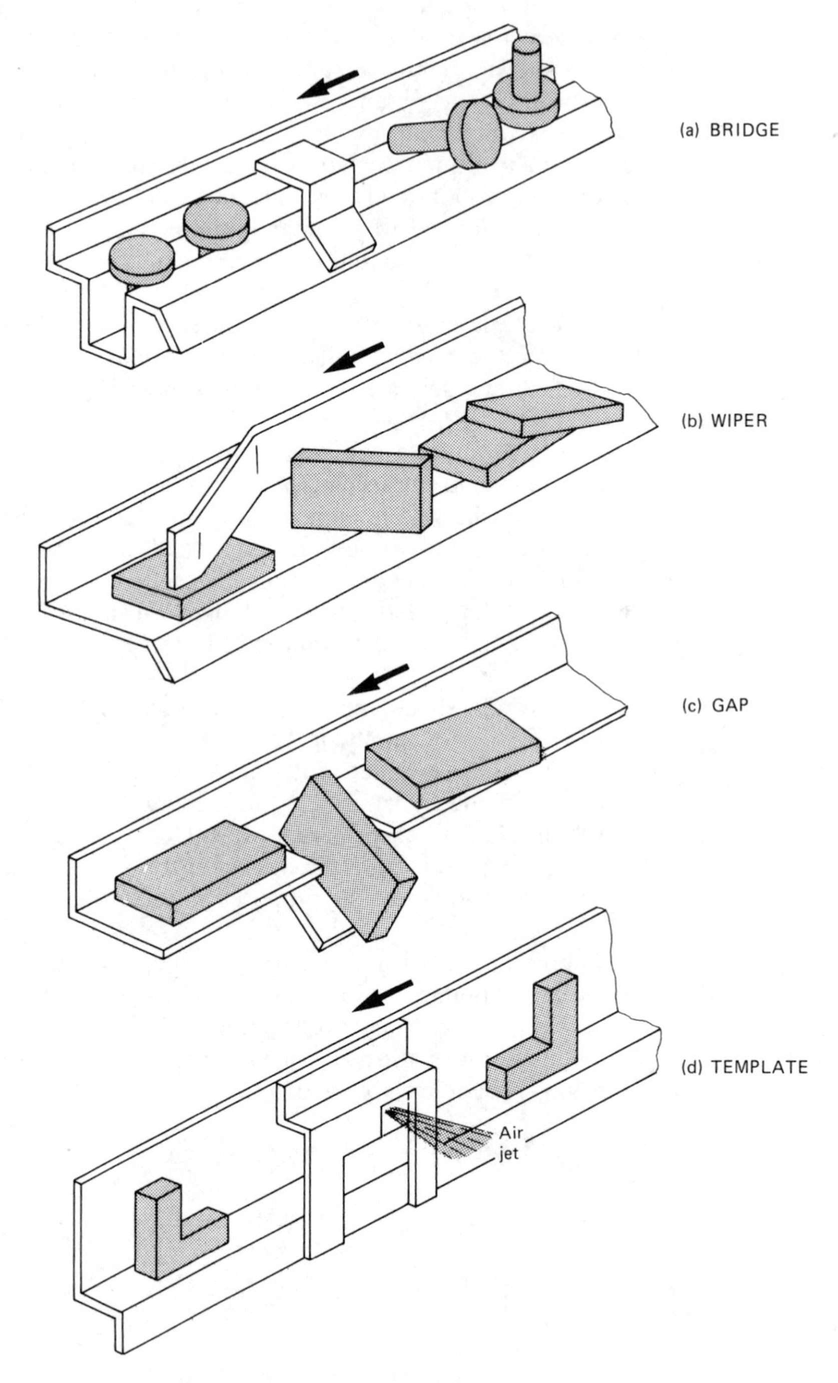

The orientation of components is usually performed by mechanical devices when the components are small. Some of these devices are illustrated in Fig. 5/17 *a*, *b*, *c* and *d*.

These are common in **vibratory-type continuous feeders**. Such feeders operate on the principle that they continuously feed components from, say, the bottom of a bowl, around a spiral path to a delivery point at the top of the bowl. On their journey upwards, the components negotiate a variety of mechanical tracks, stops, wipers, gaps and so on, designed to orientate them in a desired way. Components whose orientation happen to conform to the mechanical arrangements eventually reach the exit point in the correct orientation. Those that do not (the majority) are swept aside and dropped back into the reservoir of components for more attempts. Since the feeder operation is continuous, a steady supply of components is unceasingly fed onto the tortuous route. These devices prove remarkably efficient at maintaining a continuous supply of correctly orientated components. A vibratory bowl feeder is shown in Fig. 5/18. An obvious solution when bought-out parts are involved is to enter into discussion with the supplier to arrange that the parts are packed in a specific layout sequence and with packaging material that can be removed easily as part of the robotized unloading operation. This takes advantage of the inbuilt orientation of the supplier's packaging and possibly enables unloading and feeding to be carried out at the same time.

Fig. 5/18 Vibratory bowl feeders used in the assembly of shower bases [*courtesy: DSR Systems Ltd.*]

When the components are large, or do not lend themselves to the devices mentioned above, other methods have to be devised. **Vision systems** incorporating *pattern recognition* is an area of continuing research in the orientation of randomly presented parts. Typical approaches to pattern recognition are discussed in Chapter 6. A common means of presenting often large and irregular parts is to use manipulators. These are typically employed in fusion welding applications. A **manipulator** in this context may be described as a fixture having controlled, powered movements. It is often easier to employ a manipulator to move or rotate a component in a controlled manner, and engage the robot to traverse a relatively simple path. Thus two simply controlled movements may combine to achieve what would be a complex manoeuvre for the robot alone to perform. Consider the task of producing a circular weld around the circumference, and in the middle of, a long tube. It is often much easier to arrange that the part is slowly rotated by a manipulator, than to program a robot to reach around the stationary part. This assumes, of course, that it is convenient to rotate the component or that the component can be conveniently mounted within a manipulating device. An example of a component manipulator is shown in Fig. 5/19.

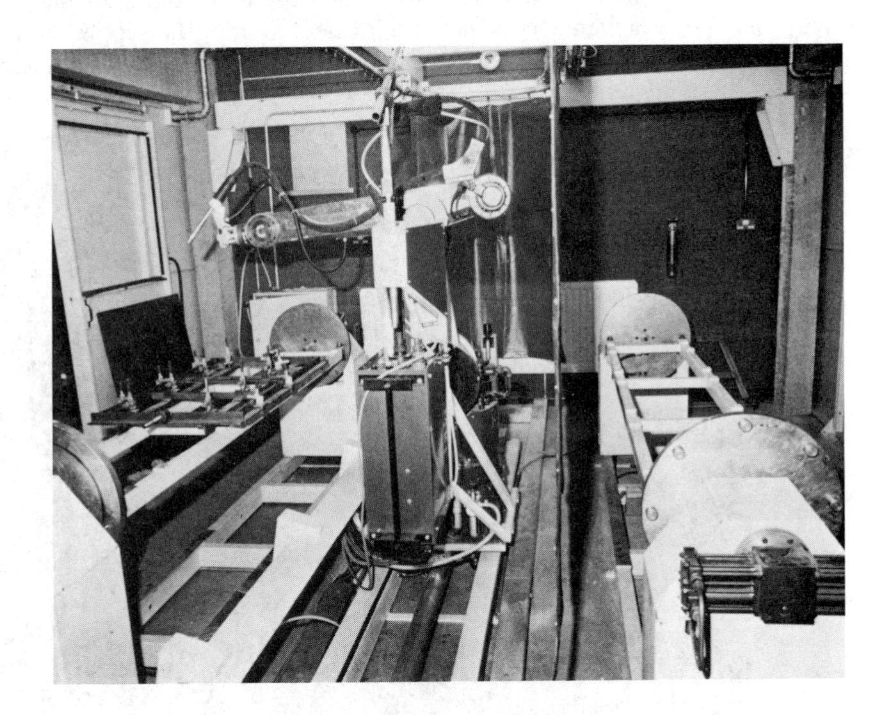

Fig. 5/19 Powered component manipulators [*courtesy: DSR Systems Ltd.*]

The **categorisation** of component forms may enable a more scientific approach to the matching of components with parts presentation methods and end effector types. Such a system may be useful in eliminating unsuitable matches rather than in attempting preferable combinations. Some possible categories of component form are

a) Cylindrical – Prismatic – Laminar + Combinations
b) Regular – Irregular + Combinations
c) Solid – Hollow – Perforated + Combinations
d) Nesting – Stacking – Interlocking – Overlapping + Combinations
e) Rigid – Semi-rigid – Non-rigid + Combinations
f) Fragile – Sensitive – Robust + Combinations
g) Combinations of the above.

Questions 5

1. Define the term 'end effector' and explain why they are not normally purchased as part of a standard industrial robot.
2. Discuss the desirable characteristics and features that should be possessed by an end effector mounting arrangement.
3. Define, with the aid of a neat sketch, the terms Pitch, Roll and Yaw in the context of end effector motions.
4. What factors should be embodied in the design of end effectors for use on industrial robots?
5. What is a multigripper and where would it be employed in a robotic task?
6. Briefly explain *three* classifications of end effectors.
7. Explain the difference between a 'hard' and a 'soft' gripper.
8. Sketch the operation of a mechanical gripper actuated by a pneumatic cylinder.
9. Explain the significance of point, line and surface contact in the context of gripper design.
10. Outline the principle of operation and the relative advantages and disadvantages of permanent and electro-magnetic grippers, stating the industrial tasks in which they could be employed.
11. Explain the principle of operation of pneumatic grippers and suggest the types of cargo for which they would be most suited.
12. Briefly explain why assembly-type operations often prove difficult to implement using industrial robots.
13. Explain, with the aid of neat diagrams, the principle of operation of a 'remote centre compliance' (RCC) end effector.
14. Identify, giving your reasons, those cargoes that would be considered difficult to deal with by industrial robots.
15. It is suggested that prior to installing an industrial robot a feasibility study should be carried out. Explain the purpose of such a study and outline the factors that must be considered.
16. Discuss how component cargoes can be 'designed for automation' with industrial robots in mind.
17. Why is 'parts presentation' important in the design of robotic handling situations and how might such parts presentation be accomplished?
18. What is a 'component manipulator', where would it be employed in a robotic task and for what main reasons?
19. Outline the use of 'vibratory bowl feeders' in the context of component handling in a robot environment.
20. Discuss the scope, benefits and limitations of employing a system of component form categorisation for use in the analysis of parts presentation for robotic handling.

Sensors and Sensory Feedback 6

6.1 Sensory feedback

6.1/0 Human senses

When human operators perform any sort of manipulative task they are constantly receiving, and acting upon, large amounts of sensory feedback. Furthermore this feedback is continually operative and does not have to be switched in and out. Individually, these senses provide information concerning the environment, the operator and the task being performed. They also operate coincidentally as safety devices through the twin mechanisms of pain and effort. More than this, the human operator is capable of combining various types of sensory feedback to perform the most complex of tasks. Illustrations of this are: hand/eye coordination in picking and positioning tasks; hand/eye/feel coordination in assembly tasks; hand/feel coordination in the qualitative assessment of surface texture, and so on. In all cases the combination of feedback information enables judgement to be exercised by the human operator. Human senses act quite automatically and for this reason are often taken very much for granted. It is worthwhile listing the types of sensory feedback enjoyed by most human operators. They include:

- *a*) Sight — including shape, size, presence, movement, distance, colour, contrast, brightness, texture.
- *b*) Hearing — including sound recognition (within a certain range), speech recognition, volume, pitch, tone.
- *c*) Smell — including aroma recognition.
- *d*) Taste — including taste recognition (salt, sweet, sour, bitter).
- *e*) Touch — including heat, force, mass, presence, texture, feel.
- *f*) Movement — including balance, direction, speed, acceleration.

In some cases human senses can be sharpened or heightened with practice and training, and they can be self-diagnostic when operation deviates from what is considered normal. However, there are also limitations with human senses that can often make them inconvenient or unreliable. Some examples include:

1 Loss of feedback can occur when human sensory organs are in some way damaged, disabled or lost completely, due, for example, to paralysis or accident.
2 Feedback dulling can occur due to the ageing process or over-exposure often causing feedback information to be converted into pain.
3 Intermittent or confused feedback can occur due to sensory overload or feedback occurring out of range of the human senses. For example, humans can only hear sounds within a certain frequency range.
4 Ambiguous feedback can occur due to the senses being 'fooled', as in the cases of optical illusions and the sensations experienced in, say, flight simulators.

Because of the above limitations there are many common devices and instruments in everyday usage that have been developed to dramatically enhance and improve the human senses. Common examples include microscopes, telescopes, X-rays, radar and so on.

It is via such sensory feedback that human operators are able to make sense of their surroundings and perform both routine and complex tasks. It also enables human operators to make sense of uncertain situations as they are encountered and adapt accordingly. It may appear (although as we shall see, not necessarily so) that if industrial robots are to be capable of performing the same types of tasks without constant human supervision then they too must be equipped with sensory feedback. Current approaches are geared to emulating the good points of many of the human senses whilst at the same time minimising their limitations. The feedback from various sensors is then analysed via digital computers and associated control software. There is a lovely human analogy of an industrial robot which imagines a deaf, dumb and blind person stood upright with feet set in concrete, with one hand tied behind and the other wearing a padded mitten. Such a being has then got to be motivated to perform accurate, repeatable and often complex tasks in the minimum time; must also be receptive to performing quite separate tasks, with quite separate sets of components in quite different environmental surroundings, at short notice; and may further have to deal with situations that can change from one execution cycle to another. This simple, almost trivial example, nonetheless highlights the extent of some of the problems involved in equipping an industrial robot to perform useful tasks in a working environment.

6.1/1 The need for sensory feedback

There are five primary reasons for employing sensory feedback in robotic devices. They are listed below. The first reason identifies a category of sensor device provided as an integral part of the physical robot and control system configuration. They are known as **internal sensors**. The subsequent reasons involve various sensor devices usually provided retrospectively either on board the robot or within the immediate working environment. They will be specifically chosen to suit the needs of the particular robot task being carried out, and the characteristics of the working environment. They are collectively known as **external sensors**, and interface to the robot control system via input/output facilities provided by the robot control system hardware and software. Different control systems are thus likely to offer quite different facilities for the connection of external sensor devices. Sensory feedback devices are required for the following reasons:

1 To provide positional and velocity information concerning the joint, arm and end effector status, position, velocity and acceleration. Usually this type of feedback is provided continuously and in real time as the robot is in motion.
2 To prevent damage to the robot itself, its surroundings and human operators by acting as safety sensors, warning or alarm indicators and cut-outs.
3 To enable the elimination of mechanically complex and expensive feeding and sorting devices.
4 To provide identification, and to indicate the presence, of different types of component.
5 To provide real time information concerning the nature of the task being performed such that the performance of the robot can become truly adaptable under varying conditions.

It is the latter category of sensory feedback that has attracted most attention with the aim of making industrial robot operation increasingly more 'intelligent'.

Arguably, the majority of industrial robots have very little built-in sensory feedback. Indeed many early control systems did not have facility for interfacing with sensor devices. There are very sound reasons for this. Industrial robots are devices that, once programmed, are extremely good at performing predictable and precisely defined tasks. The accuracy and repeatability with which they perform various tasks, and the suitability of a particular robot configuration for a particular task, will be a function of its design, construction and control system. In just the same way, the human analogy illustrated above could perform similar precisely defined tasks within the constraints of its own particular physical limits and working environment. In these cases there is little need for intervention by sensor devices and therefore no real need to rely on sensory feedback. However, take away the preciseness and predictability of the task and an industrial robot is likely to become a liability unable to perform even the most basic of tasks. As the tasks to which industrial robots are applied become increasingly more demanding, the **variability** with which they have to deal also increases. Thus, there is an increasing need for sensory feedback.

Sensory feedback is needed whenever there is uncertainty, unpredictability or imprecision present in any aspect of the tasks being performed. There is a very important lesson to be learned here in that, if this uncertainty, unpredictability and imprecision can be designed out of a task, then the need for sensory feedback is often eliminated. There are many illustrations of this. For example, in continuous fusion welding applications it is common to employ joint or seam sensors to enable to robot to accurately follow the desired weld line. One reason for this may be that there is variability in the fit of the parts to be joined, caused perhaps by the way in which the parts themselves are manufactured. If the variability of fit can be eliminated, for example by tightening up the manufacturing process that produces the components, then the need for joint-following sensors could be eliminated. A second example concerns an industrial robot equipped with sensors to identify the position and orientation of previously machined components, unloaded from a machine tool and randomly placed on a storage rack. Since the components have perfect position and orientation whilst they are still located in the machine tool, it would simplify procedures if the robot could be arranged to handle the components directly from the machine. This makes more sense than discarding the position and orientation by unloading from the machine and then trying to recover the same situation by complex sensing techniques.

These examples serve to illustrate the thought processes that should be considered prior to opting for sensory feedback solutions. The advantages may extend beyond the possible elimination of expensive sensor devices. For example, a cheaper or simpler robot/control system configuration could possibly be specified, in which case programming time and effort will be simplified, cycle times will be reduced and the task should be more reliable since a potential weak link in the chain is eliminated. Conversely, it may make economic sense to employ rudimentary sensory feedback rather than leave trivial, simple or mindless tasks to human operators.

6.1/2 Levels of variability

Research into sensors and sensory feedback for use in providing robot adaptability has identified three levels at which sensors can counter variability or uncertainty in industrial tasks. For each of the three levels a different strategy has evolved for its solution, assuming that sensors have to be employed.

Level 1

Variability — Different components vary, in dimension, form, position or orientation from one cycle to another.

Strategy — Each component is probed or sensed once, prior to the execution of the task, as part of the operation cycle. Adjustment of the program (or its parameters) is made prior to performing the task.

Level 2

Variability — Different operations within the same overall task may incur different levels of variation. For example, a number of different, randomly placed components may have to be negotiated at different stages of an assembly-type task.

Strategy — Sensory intervention and compensatory adjustment must be available, within the program, at each discrete stage of the task.

Level 3

Variability — Uncertain changes are likely to occur during execution of the task.

Strategy — Continuous sensing and feedback, in real time, must be provided throughout the complete execution of the task.

Employing sensory feedback also gives the potential of increasing the stated accuracy and repeatability specifications of a particular robot.

6.2 Sensors

6.2/0 Analog and digital signals

In any electronic system two types of electrical signal may be encountered. In Chapter 4 the concept of a digital signal was introduced. A **digital signal** is either a voltage ON or a voltage OFF, allowing two discrete states to be detected. More states may be detected by combining, and interpreting, the ON/OFF conditions of a number of different digital signals taken together. The voltage levels used are normally 5 V (representing the ON state), and the 0 V (representing the OFF state).

The second type of signal is represented by a varying voltage that is allowed to take up any level in between a maximum and a minimum value. This type of signal is termed an **analog signal** or an analog voltage. Analog voltages are characterised by the way they can change continuously over even short periods of time. There are no accepted high or low standard values for analog voltages. Wherever they are used with digital computer systems, however, they will be designed to operate within the range of the digital voltage levels of the computer system. This will probably be achieved by conditioning the signals through some sort of interface. Examples of common analog quantities (i.e. quantities which change continuously over time) that may need sensing in robotic applications include voltage, temperature, light, mechanical displacement, velocity and acceleration.

It should be reasonably clear that, even if analog and digital signals are both within the same voltage range (say 0 V and 5 V), they are fundamentally incompatible with each other. Since a digital signal is simply ON or OFF there is no way in which it can express the gradation of an analog signal. In many practical robotic situations, digital control systems are required to interface with analog systems. For example, a command instruction from the robot controller may signal that an axis motor should operate to vary the position of the robot arm. If the axis drive is, for example, a DC motor then it will require the application of an analog (varying) voltage. Similarly, as the robot arm moves, its position will change continuously. This continuous (analog) change in displacement must be fed back into the digital robot control system such that the controller can monitor the exact position of the robot arm. It follows that if the two types of information are to be transferable between systems then some means of signal conversion must be accomplished.

The techniques of converting signals from one type to the other are referred to as *digital-to-analog conversion* (DAC) and *analog-to-digital conversion* (ADC). Although there are a number of ways of arranging signal conversion, the most convenient is to utilise one of the readily available, purpose-built integrated circuits (ICs). These IC chips are known as D-to-A or A-to-D convertors. **D-to-A convertors** are devices that accept input in digital form and give a varying analog voltage output proportional to the binary value input. They are available for performing conversion on 4, 8, 10, 12, 14 and 16-bit binary inputs, acceptable in either serial or parallel input modes. **A-to-D convertors** accept a varying analog voltage input and supply a proportional digital value as output. Because analog voltages can vary over very small periods of time, the speed of operation of A-to-D convertors becomes an important factor. A-to-D convertors get progressively more expensive as the speed of conversion (termed the *conversion time*) increases. Relatively slow conversion times are measured in milliseconds (thousandths of a second), whilst relatively fast conversion times are measured in microseconds (millionths of a second).

Exercise Identify those points within a particular industrial robot/controller configuration where analog and digital voltages are likely to be present, and where A-to-D or D-to-A conversion is required.

6.2/1 Transducers and sensors

The **transducer** was introduced in Chapter 3. It is a general term defined as any device that converts energy in one form into energy in another form in such a way that the output is proportional to the input. The output from most transducers is in the form of electrical energy. This is most convenient since electrical voltages are easy to sense, easy to measure and easy to manipulate. Voltage signals can easily be stepped up (increased) or stepped down (decreased), or converted into other signal types. Examples of common transducers include motors (electrical energy into rotary mechanical energy), dynamos (mechanical energy into electrical energy), and solenoids (electrical energy into linear mechanical energy). Transducers need not be discrete components in themselves but may be designed from a number of separate component parts.

Sensors differ slightly from transducers in that, although changes occur, the changes do not necessarily involve energy conversion. It is more likely that a change in some property of the sensor occurs. For example, many of the position measuring transducers introduced in Chapter 3 utilise a component called a photocell. A photocell is a device that changes its electrical resistance in proportion to the amount of light falling on it. If a voltage is connected across the photocell then this voltage changes as the amount of light changes. Thus, the photocell can be caused to act like a transducer although strictly speaking no energy conversion is taking place. Sensors are more likely to be discrete components in their own right and many are 'packaged' for use in certain environments (e.g. heat or pressure resistant). The difference between a sensor and a transducer, however, is not of great significance and the remainder of this chapter will not distinguish between them. Sensing devices are probably best considered from a systems (black box) point of view. It is generally only necessary to know the characteristics of the output of the sensor for a given input.

It is necessary to exercise care in the choice of sensing devices for particular tasks. The operating characteristics of each device should be closely matched to the task for which it is being utilised. Different sensors can be used in different ways to sense the same conditions, and the same sensors can be used in different ways to sense different conditions. Five important characteristics of any sensing device are:

a) **Range** The range refers to the maximum and minimum change in input signal to which the sensor can respond.

b) **Response** The sensor must be able to respond adequately to the smallest incremental change that is likely to be significant for the application concerned. It may also be termed **frequency response**.

c) **Accuracy** The output of the device should properly reflect the input quantity being measured or sensed. There should be facility for calibrating the device to ensure faithfulness of reproduction.

d) **Sensitivity** Sensitivity refers to the change in output exhibited by the sensor for a unit change in input.

e) **Linearity** For a device to be linear it must exhibit the same sensitivity over its complete measuring range.

Other considerations include:

a) Obviously important are the physical size of the sensor and where it is to be utilised. Access may be important for its adjustment or replacement.
b) The device should itself not disturb or have any effect upon the quantity it is to sense or measure.
c) The device should be suitable for the environment in which it is to be employed. For example, it should be robust enough to withstand any harsh physical treatment or hostile environmental conditions (such as heat) that it is likely to encounter. It may be necessary to provide physical protection.
d) Ideally, the devices should include isolation from receiving (or transmitting) excess signals or electrical noise that could cause overloads, giving rise to the possibility of sensor, circuit or system mis-operation or damage.

The deployment of sensing devices in the workplace must be justified in either economic or technical terms. It may be that, by slight re-organisation of the workplace, or changes in the operation sequence, component design or current practices and procedures, the need for sensors can be eliminated.

6.2/2 Sensing devices and applications

There are a number of ways in which sensing devices may be broadly classified:

a) By their type of operation—analog or digital.
b) Whether the quantity or attribute is sensed directly or indirectly.
c) By the medium by which they operate—optical, electrical, etc.
d) By their application.

Since the same devices can perform quite different tasks in different circumstances, the devices and techniques will be considered, in this section, according to the common applications for which they are employed.

1 PROXIMITY SENSING

Proximity sensing normally means detecting:

a) either the presence or absence of an object
b) or the size or simple shape of an object.

Proximity sensors can be further classified as contact or non-contact, and as analog or digital in operation.

Detecting the presence or absence of a component is a fairly simple process. Basically all that is required is a two-state signal (a switch) to indicate either presence or absence. This is essentially a digital operation and may be accomplished in a number of ways. The choice of sensor devices will be determined by physical, environmental and control considerations and the attributes and characteristics of the sensors available. They include the following;

Mechanical Any suitable mechanical/electrical switch may be adopted but, because a certain amount of force is required to activate a mechanical switch, it is common to settle for **microswitches**. These are miniature switches requiring little force to activate them. They are often convenient since they are small, discrete, easy to mount and can be adapted to a number of situations by attaching suitable lever arms to operate them. They are a contact device and care must be exercised to ensure that any contact with the switch does not displace or impede light cargoes or components. They are easily integrated into control systems since they are electrical in operation. Electrical connection can usually be made such that the switch either makes the contact to allow the signal to pass (termed *normally open*), or breaks the contact to discontinue a previously applied signal (termed *normally closed*). Thus, in digital terms, both a 1 to 0 binary transition or a 0 to 1 binary transition can be detected.

Pneumatic Pneumatic proximity sensors operate by breaking or disturbing an air flow. The mode of sensing may be to remove air pressure from a sensing device (by breaking the air flow), by creating a back pressure in the supply (by disturbing or impeding the air flow), or by releasing pressure from a pressurised system. Low-pressure air supplies are normally used in conjunction with diaphragm-type switches. Obviously they cannot be employed unless there is a convenient air supply available and additional equipment is likely to be required to provide a clean supply and minimise fluctuations in air pressure. Although the sensing device itself may not make contact with the component, the pressurised air medium probably will. They can thus be thought of as contact-type sensors. Pneumatic sensors cannot therefore be used where there is a likelihood that light components will be blown away or deviated by the air flow. Simple pneumatic safety devices can be constructed from tightly coiled springs through which compressed air flows. In the case of a collision causing the spring to deflect, loss of air pressure will occur due to the air escaping through the expanded and separated coils to atmosphere.

Optical In their simplest form, optical proximity sensors operate by breaking a light beam that impinges directly, or is reflected, onto a light-sensitive device such as a photocell. Optical sensors are examples of non-contact sensors. Care may have to be exercised in environments where the ambient light level is subjected to momentary disturbances. For example, optical sensors could become 'blinded' by flashes from arc welding processes, airborne dust and smoke clouds may impede light transmission and so on. Light beams can easily be conditioned to operate on small or large areas using lenses, and in obstructed areas via mirrors and optical fibres. In such cases a clean environment may be essential to maintain the integrity of the optical devices. Other devices like infra-red transmitters and receivers are also a form of optical sensor. Devices can also be made sensitive to certain colours (and heat).

Electrical Electrical proximity sensors may be contact or non-contact in operation. Simple contact sensors can be made by making the sensor and the component complete an electrical circuit causing a current to flow. The obvious limitation of this technique is that the component and the sensor must be electrical conductors. Non-contact electrical sensors rely on the electrical principles of either inductance for detecting metals, or capacitance for detecting non-metals as well.

Some proximity sensors make use of a combination of the above techniques within a single sensing device. For example, an opto-mechanical sensor can cause a light beam to be interrupted by the mechanical operation of a switch acting as a shutter. By employing two or more of the above sensors in conjunction with each other, different sized components, rudimentary shapes of different components or simple orientation of components can be determined. These techniques are further discussed under tactile sensing below. Ultrasonic and infra-red transmitters and receivers can also act as proximity and intrusion sensors. Some examples of sensor devices are illustrated in Fig. 6/1.

2 RANGE SENSING

Range sensing concerns detecting how near or how distant a component is from the sensing position, although they can also be used as proximity sensors. Distance or ranging sensors use non-contact analog techniques. Short-range sensing may be accomplished with electrical capacitance, inductance and magnetic techniques. These can be made active within a range of a few millimetres to a few hundred millimetres. Longer-range sensing usually involves transmitting energy waves of various types. The transmitter can be set up to aim its energy wave at a receiver situated some distance away. The time taken for the energy wave to reach the transmitter is a measure of the distance between them. This is likely to be more successful when the transmitted wave is discontinuous.

Continuous ranging can be achieved by transmitting energy waves towards target components and detecting the time taken for the waves to be reflected back. This principle is well established in radar (using radio waves) and sonar (using sound waves) techniques. The use of lasers for range sensing is also being researched. Considerable sophistication may be required in translating and interpreting reflected information since the waves are likely to be reflected by both the target components and other objects within the path of the energy wave. Extremely long-distance ranging can be accomplished using these techniques. Ranging sensors may also be employed as a form of proximity sensor to seek and find various target objects as part of guidance systems.

3 FORCE SENSING

There are six types of force that may require sensing. In each case the application of the force may be static (stationary) or dynamic (moving). Force

PROXIMITY SENSORS

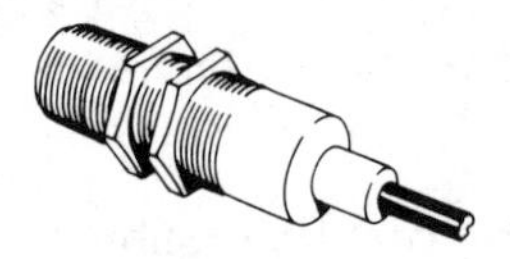

Capacitive

A range of proximity detectors with built-in amplifiers and solid-state output stages making them extremely versatile electronic switches. These sense the presence of non-conducting materials such as wood, PVC, glass, etc., as well as ferrous and non-ferrous metals. All types have built-in potentiometers for sensitivity adjustment and LED indicators. Applications include batch counting, alarm systems, limit switching, etc.

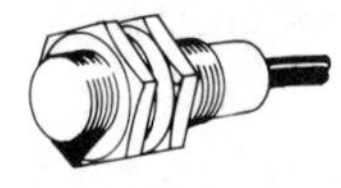

Inductive

A miniature inductive proximity detector housed in a threaded anodised aluminium cylindrical case with integral 3-core PVC sheathed cable. Environmental protection to IP 67. The device incorporates electrical protection against reverse polarity, supply line and load transients and has a current limiting PTC resistor in the load output.

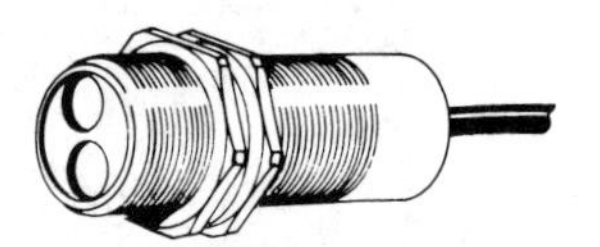

Optical

A high-quality proximity switch operating on the principle of the emission/detection of modulated infra-red light. This mechanically and electrically rugged unit is housed in a threaded aluminium case with a glass end window and has many applications where any material is required to be detected, at a range far exceeding that associated with conventional proximity devices.

GAS SENSOR

A platinum wire (pellistor) flammable gas sensor that is designed to sense such gases as propane, butane, methane, isobutane, liquefied petroleum gas, natural gas and town gas. The performance of the sensor is stable against change of ambient temperature and humidity. Bridge supply voltage 3 V d.c. + 10%.

ULTRASONIC TRANSDUCER

An ultrasonic transmitter and high sensitivity receiver designed for sending and receiving continuous or modulated waves in the 40kHz region through air. Applications include remote-control systems, batch counting, data transmissions, etc. Typical operating distance 5m. **Note:** These units are not watertight.

STRAIN GAUGE

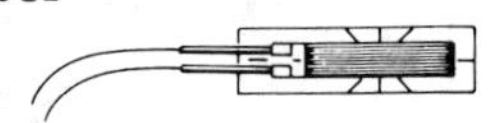

General-purpose foil-type polyester-backed strain gauges. Available with temperature compensation for steel or aluminium.

SLOTTED OPTO-SWITCH

A slotted opto-switch comprising an infra-red source and integrated photodetector. The i.c. photodetector consists of a photodiode, amplifier, voltage regulator, schmitt trigger and output stage. An important feature is it wide supply voltage range, +4.5V to 16V, which allows the output to interface directly to TTL, LS/TTL and C-MOS. The detector can sink 10 TTL loads when the infra-red beam is interrupted. The output rise and fall times are independent of object speed and the integrated voltage regulator ensures high noise immunity. Applications include position detection, paper sensing, counting, optical encoding and level sensing. Operating temperature range: -40°C to +100°C.

REFLECTIVE OPTO-SWITCH

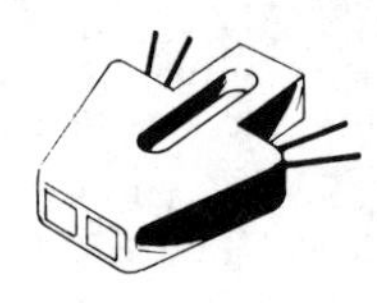

An infra-red emitting LED and phototransistor sensor housed in a moulded package. The phototransistor responds to radiation from the diode when a reflective object is placed within the field of view. The device is ideal for counting shaft revolutions, line counting on paper tapes, etc. The moulded case incorporates a dust cover and infra-red filter to prevent the ingress of dust and eliminate ambient illumination problems. A slotted mounting hole allows for adjustment to the sensing distance.

Fig. 6/1 Typical sensing devices

is a vector quantity in that it must be specified by both a magnitude (size) and a direction. Force sensors are therefore analog in operation and sensitive to the direction in which they act. The six types of force are:

1. Tensile force (pulling)
2. Compressive force (pushing)
3. Shear force (sliding or tearing)
4. Torsional force (twisting)
5. Bending force (bending)
6. Frictional force (gripping)

A number of techniques exist for sensing force, some direct and some indirect via changes in other quantities. The techniques and devices selected will be dependent on the type and magnitude of the force and the mode of application of the force.

Tensile forces can be determined by devices called **strain gauges**. These are small devices that exhibit a change in their electrical resistance when their length is increased. Strain gauges are affixed to the components being sensed, usually by adhesives, in the direction of the applied tensile force. The change in electrical resistance of the strain gauges can be translated into force, and as such these are indirect devices.

Compressive forces can be determined by devices known as **load cells**. These operate by detecting a change in dimension of the load cell under the application of a compressive load, or by detecting an increase in pressure within the load cell due to the applied load, or by exhibiting a change in electrical resistance under the influence of a compressive load. The latter may be termed *piezoresistive sensors*.

Torsional and bending forces can be visualised as incorporating directionalised compressive and tensile forces. As such, combinations of the above techniques and devices can be employed to determine torsional and bending forces.

Frictional forces usually relate to dynamic situations or where movement is to be restrained. Frictional force may thus be detected indirectly using a combination of force and movement sensors. This approach is illustrated in Fig. 6/2 to sense and control the gripping pressure in an end effector gripper.

Fig. 6/2 Sensing of gripper pressure

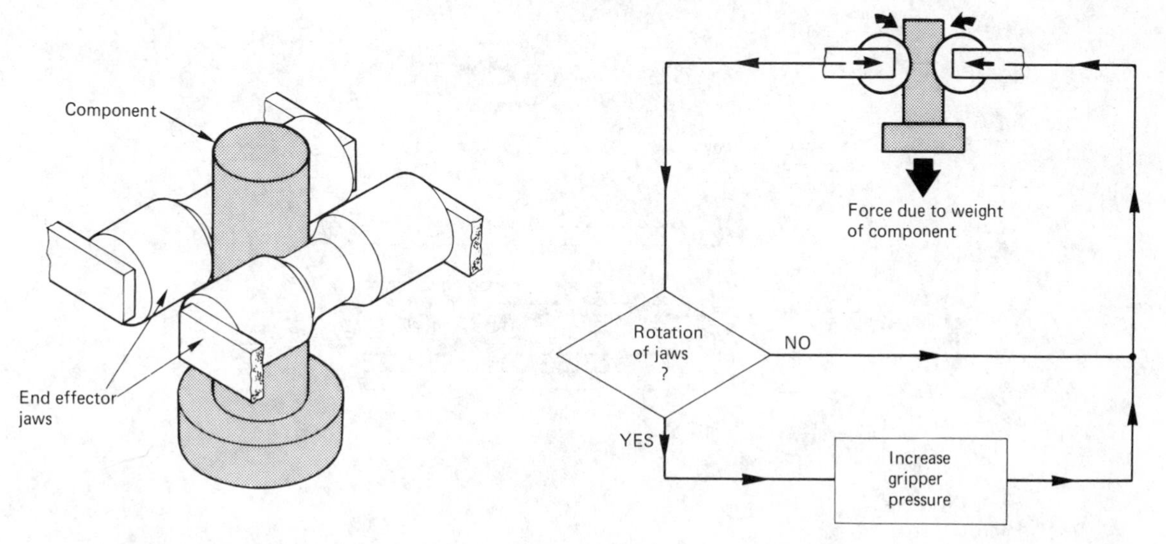

4 TACTILE SENSING

Tactile sensing means sensing through touch. In its widest context tactile sensing can be taken to mean any sensing where contact is involved, for example a form of contact proximity sensing using microswitches. Since proximity sensing has already been discussed, the following discussion adopts the term tactile sensing to mean the sensing of component shape or size by direct contact. The simplest mode of application is to provide a number of simple touch sensors in the form of a square or rectangular array, where the sensors are arranged in rows and columns. They are commonly called **matrix sensors**. Each individual sensor is activated or not when brought into contact with the object. By detecting which sensors are active (in the case of digital sensors), or by analysing the magnitude of the output signals of the sensors (in the case of analog sensors), the imprint of the component being touched can be determined. This imprint is then compared, under software control, with previously stored imprint information to determine the size or shape of the component. Mechanical, optical and electrical tactile sensors are readily available. Such sensors are commonly covered with a sensitive gossamer rubber or polyurethane skin as a form of protection and to smooth out the readings obtained. The principle of a tactile sensor is illustrated in Fig. 6/3.

Fig. 6/3 Principle of a tactile matrix sensor

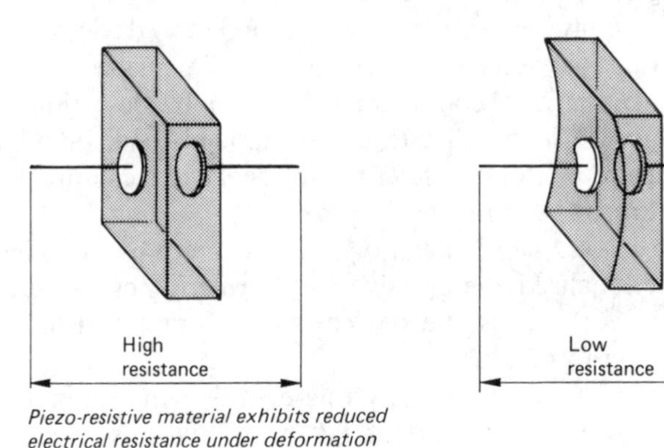

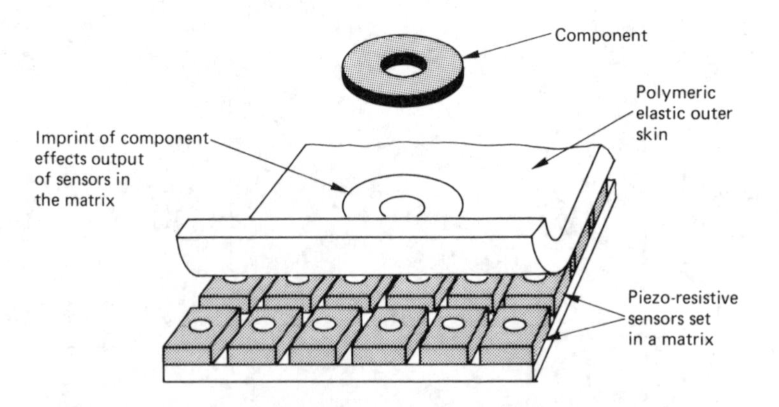

5 HEAT SENSING

Heat sensing may be required as part of process control or as a means of safety control. A number of techniques can be applied, the choice being determined to a large extent by the magnitude of the temperature to be sensed.

The simplest type of mechanical binary heat sensor is the familiar *bi-metallic strip* employed in many thermostat devices. Two differing materials are joined together in the form of a composite strip. The materials are chosen such that their coefficients of linear expansion differ. As the sensor is exposed to a heat source, the metals expand at a different rate thus causing the strip to bend. The bending of the strip can be caused to operate a pair of contacts to signal a certain temperature threshold. These devices are often used as safety cut-outs since, when the heat source is removed, the metals contract and the sensor reverts to its original dormant condition.

Analog heat sensors operate as thermometer-type devices. They operate according to the change of some quantity in response to the application of heat. For example, *electrical resistance thermometers* utilise the change in electrical resistance of a fine platinum element (−50 to 500° C). Metallic conductors all exhibit an increase in electrical resistance with an increase in temperature. *Thermocouples* are transducers that consist of a pair of dissimilar metals welded together at one end. Application of a heat source causes a small voltage to be generated in the wires that is proportional to the amount of heat applied (1000° +). *Thermistors* are made from non-metallic electrical components called semiconductors. These devices exhibit either a decrease in electrical resistance as temperature increases or an increase in electrical resistance as temperature increases, depending on their make-up. They are employed when a fast response to small changes in temperature is required. *Infra-red cameras* represent sophisticated systems for detecting low-level heat sources.

6 ACOUSTIC SENSING

Acoustic sensors detect and sometimes discriminate between different sounds. They can be employed in speech recognition systems where spoken commands can be recognised, or in situations where abnormal sounds such as explosions need to be detected. The most common acoustic sensor is the familiar *microphone*. This device enables an analog voltage to be generated in response to sound vibrations. The resulting analog voltage trace can then be compared with some (previously learnt) reference trace in order to achieve discrimination. The obvious problem with acoustic sensing in an industrial environment is the high incidence of background noise likely to be present during the course of normal working practices. Acoustic sensors can easily be tuned to respond only to certain sound frequencies, thus making them discriminatory.

7 GAS SENSING

Gas or smoke sensors which are sensitive to particular gases rely on chemical changes in materials contained within the sensor devices. Chemical changes and reactions may produce physical expansion or heat sufficient to trigger a switching device.

Vision sensing and pattern recognition sensing are dealt with in Section 6.3.

6.2/3 Control of sensory feedback

Sensory feedback is supplied to the robot control system via input ports provided in the design of the computerised control unit. Such input is ultimately in digital form. The control unit itself may provide facilities for analog-to-digital conversion or it may be carried out, as required, by external interfaces. (Refer to Section 4.3/4 for further information on how this is accomplished.) However, providing physical connections (hardware) for the various feedback signals is insufficient in itself to take advantage of the incoming information. The main robot control operating system (software) must also be configured to recognise when feedback information is present, and be able to respond in an appropriate manner. Feedback signals can be awkward to deal with since they can occur at any time during the execution of the operating system software. The operating system must also simultaneously and continuously attend to the many other tasks required to ensure normal robot operation and function, when feedback information is not present or does not require service.

Within the control system software, assuming that the occurrence of an input signal can be detected, there must be some means of servicing the input signals. This will probably take the form of purpose-written *subroutines*, called whenever an input signal is recognised. Such subroutines are often called **service routines**. Different feedback signals require, and are handled by, their own service routine.

There are two basic software concepts available for dealing with external information fed into a computerised system. They are termed *polling* and using *interrupts*.

Polling

When polling, the operating system software systematically tests, as part of its routine tasks, each input port to see if there is an input signal that requires service. If there is a feedback signal then the program branches to a section of program code (the service routine) that analyses the signal and takes the appropriate action. After the input has been serviced, program control is returned to the polling process. If there are several input ports then two disadvantages present themselves. First, a great deal of time will be spent in polling the ports even when no inputs are present. This means that the microprocessor and the operating system will be needlessly tied up in just examining the status of the input ports. Secondly, since the ports are examined sequentially one after the other, a delay occurs in between each examination of the port. This could result in input data being lost or, more commonly, a device receiving service too late. The advantages of using polling techniques are that they are simple in operation and that no hardware assistance is required to deal with the feedback signals.

Interrupts

A special hardware line, called the interrupt line, is provided via a special pin of the microprocessor. When this line is pulsed with a digital signal the microprocessor, under the control of the operating system, acts as follows:

First, it completes executing the instruction it is currently processing.
Secondly, it saves the position within the program it is currently executing in a specially reserved part of memory (called the *stack*).
Thirdly, it saves its own status condition (also on the stack).
Finally, it branches to an *interrupt service routine* to handle the interrupt and process the available data.

Thus, all external sources of feedback signals are configured to pulse the interrupt line when they require service. On the completion of the service routine the microprocessor recovers its status (i.e. before it branched to the service routine), and resumes execution of the program from the point at which it was interrupted. The facility for external devices to interrupt the processor can be switched in or out under software control. This enables the system to fully complete any instructions that have a higher priority than external information, without fear of being interrupted. There also needs to be facility for the 'stacking' of multiple interrupts that could occur from different sources, even during an interrupt service itself. In practice, when using interrupt techniques, only a small number of devices are likely to be connected at any one time.

The obvious advantage of interrupt techniques is that the processor does not have to waste time polling barren input ports. The disadvantages are that the programming is more difficult and that hardware support is required.

6.3 Robot vision

6.3/0 The scope of industrial robot vision

Perhaps the most active field of current research into sensory feedback concerns robot vision. **Robot vision** means the capture of an image in real time via some form of camera system, and its conversion into a form that can be fed into (and analysed by) a computer system. This conversion process is termed **digitising the image**. The entire process (image capture, digitising, and data analysis) should be quick enough to enable the robotic system to respond to the analysed image, and take appropriate action, during the performance of its task. The perfection of robot vision techniques will enable the full potential of artificial intelligence to be bestowed on industrial robots. Its uses include the detection of presence, position and movement; the recognition and identification of different components, patterns and features; and complex scene analysis and identification applications.

We shall see, however, that even the most rudimentary vision techniques of data capture and analysis are extremely greedy of computer memory and can take relatively long processing times, especially if a poor image has to be enhanced by computer techniques. For these reasons robot vision systems are usually monochrome (black and white) and many incorporate dedicated microprocessor-based computer systems to support their image processing activities before handing over the captured (and possibly analysed) data to the robot control system. This is termed **preprocessing**. The vision camera itself may be located at a fixed, strategic location within the workplace, attached to the robot structure or incorporated into the robot end effector. Developments in fibre optic techniques open up possibilities of viewing images, and 'placing' a lens into restricted areas to which conventional cameras could not normally gain access.

Industrial robot vision systems can often be difficult to implement since they are subject to many external influences such as: movement and vibration; momentary obstructions; ambient lighting levels; shadows; reflections and glare; component/background contrasts; and dust, smoke and other airborne particles that could distort, degrade or confuse the images produced. These important influences may be difficult to control and maintain successfully within a busy industrial production manufacturing environment.

6.3/1 Elements of robot vision systems

In any successful industrial robot vision system four fundamental elements must be present. These are:

A source of illumination	(to create correct lighting conditions)
A camera system	(to create and focus a suitable image)
A computer interface	(to digitise and store the image)
Appropriate software	(to analyse the image)

1 Source of illumination

A suitable source of illumination is required for two primary reasons. First, for any photographic or vision system to produce a suitable image a minimum amount of light is required. For video camera techniques, the level of light required is usually quite high. The ambient lighting conditions of a factory workplace incorporating large casts of shadow from stationary machine tools, moving robots, personnel and material handling equipment, and the differing levels of light shining through outside windows (depending on the time of day and weather conditions), cannot be relied upon to provide consistent lighting conditions.

Secondly, the ability to differentiate between a subject and its background depends mainly on the level of contrast within the image. Obtaining stark contrasts means arranging lighting conditions such that the background is uniformly dark whilst the subject(s) stand out uniformly in light, or vice versa. Image-enhancing techniques can be used which convert progressive shades of grey into either light or dark, thus increasing the contrast between the subject and its background. The creation of stark contrast is for the benefit of the software that has to identify subject from background when processing image data. Since the subject's material, colour and texture will usually be fixed, these conditions can be achieved by suitable choice of background material and the mode and direction of the light source. Dark background and light subject, for example, is best achieved by lighting the subject from the front, whilst a light background and dark subject is better achieved using backlit methods. The placing of the light source is also an important influence since its direction can cause the subject itself to cast shadows. Unless the vision system is very sophisticated, it will not be able to distinguish between the subject and its shadow and a confused image will result. In addition, the light should be diffused (evened out) to avoid light concentrations (hot spots) falling onto the subject. Hot spots can suffer the same misinterpretation as shadows.

2 Camera system

Vision sensing cameras are basically of two different types. The first type incorporates a conventional **TV camera tube**, sometimes called a vidicon tube. These work by forming an image on a photosensitive screen and quickly scanning the image electronically from top to bottom. The result of the scanning process is an analog signal that continuously varies with the position and light intensity of the screen as it is scanned. The second type is a **solid state camera** (electronic with no moving parts) which consists of either a row (linear array) or a matrix (two-dimensional array) of photosensitive cells. The individual cells become charged with a voltage the size of which depends on the amount of light falling on them. They are often termed *charge coupled devices* (CCD). The individual rows of electrical charge can be electronically shifted sequentially, detected and combined to produce a continually changing analog voltage which represents the image scanned. The CCD devices represent the more modern technology and are likely to increasingly displace vidicon tubes for robot vision applications, for the following reasons:

a) They cannot be permanently damaged by exposure to intense light.
b) They do not suffer from image smearing due to the scanning process.
c) Being solid state they consume relatively little power.

Different lenses may be considered to limit the depth of field, which is the depth of image that will be in focus. This becomes more important in environments with busy backgrounds where only the foreground object of interest needs to be in focus. The mounting of the camera may be important to minimise the effects of parallax. Parallax is the distortion observed when objects are viewed at an angle rather than normal (at 90°) to their profile. The effect is easily demonstrated. A coin is circular in shape when viewed, at eye level, at right angles to its face. If the coin is tilted (say at 45°) away from the line of sight without altering its position then the shape assumes that of an ellipse. Taken to the extreme, if the coin is further tilted (through a full 90°), the shape becomes a straight line or a thin rectangle.

3 Computer interface

A computer interface is required to link the output from the camera system to the computer system that will carry out the image analysis. The interface may be required to carry out a number of tasks. Since the output from both the above camera types is an analog voltage, then analog-to-digital conversion of the signal must be carried out to enable it to be input to a digital computer. This process is termed *digitising*. The analog voltage video signal may need filtering of any interference (termed *noise*) it picks up during image capture and signal transmission. The image may also need enhancing to improve subject/background definition and contrast. Finally, the image may undergo processing or analysis before onward transmission to the computer system. Because of the speed at which image data is sent by the camera system, it is likely that such an interface would utilise dedicated high-speed processors.

4 Computer software

Once a suitable image has been digitised, it then has to be interrogated to extract meaningful and relevant information. In order to do this a strategy or set of rules (called an *algorithm*) has to be established to determine the required features of the image. The rules, in turn, have to be converted into a program (the software) that will carry out the analysis and provide results in the form of data which can be used by the robot controller to direct the actions of the robot accordingly. In order to reduce processing time this program code should be as short and as efficient as possible. There are a number of types of information that may be obtained from a digitised image and these will be discussed in Section 6.3/3.

6.3/2 Principles of image representation

The analog signal obtained by the vision camera, after processing through an A-to-D convertor, is eventually converted into a digital representation of the captured image. A digital representation of the image means that the image information (in terms of its lightness and darkness) is stored sequentially as binary bit patterns in computer memory. The information contained in a single memory location (a byte) can thus be used in different ways depending on the vision system operation.

At its simplest level each individual bit can be interpreted as either light (when set to binary 1) or dark (when set to binary 0). This has the disadvantage that only a coarse white/black computer image (with no gradation of contrast) will be generated from the camera image. The vision system will have some means of setting a threshold level at which light levels are deemed to be light or dark. The quality and integrity of the image needs to be interpreted with much caution. It will, for example, be sensitive to lighting conditions. Any shadows will probably be decoded (and hence digitised as black areas), whilst areas of reflection and glare wil be digitised as white areas. In addition, no depth of form on the component can be detected and hence this limits images to simple two-dimensional silhouettes. Other problems, such as the overlapping of parts (called *occlusions*), which are easily detected with the human eye, cannot easily be dealt with by this type of system. The silhouettes can become significantly enlarged, reduced or distorted unless ideal conditions are present. This situation is quite unacceptable since it must be understood that, even when an image has been digitised and appears on a VDU screen, it is not the human operator that has to interpret and make sense of the image, but the computer. This has to be done entirely by a methodical method of interrogating the contents of memory by a software program. The advantage with this system of interpretation is that it makes optimum use of computer memory since eight picture elements can be represented by one byte of memory. One individual picture element is termed a **pixel**. A low-resolution digitised image utilising 64 × 64 pixels can be accommodated in 64 bytes of memory. Two small overlapping washers digitised by such a system are shown in Fig. 6/4. In addition to illustrating the principle it also highlights the obvious limitations of such a rudimentary system.

Because of the above limitations a second approach enables gradation of contrast to be identified. Reference back to Chapter 4 will confirm that 8 bits

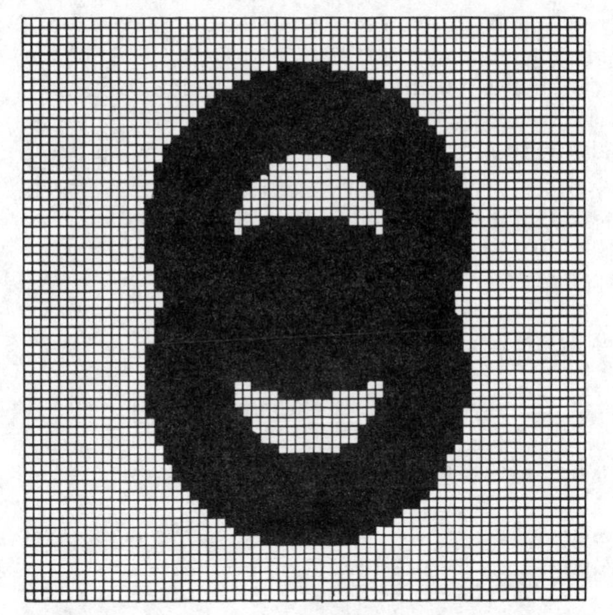

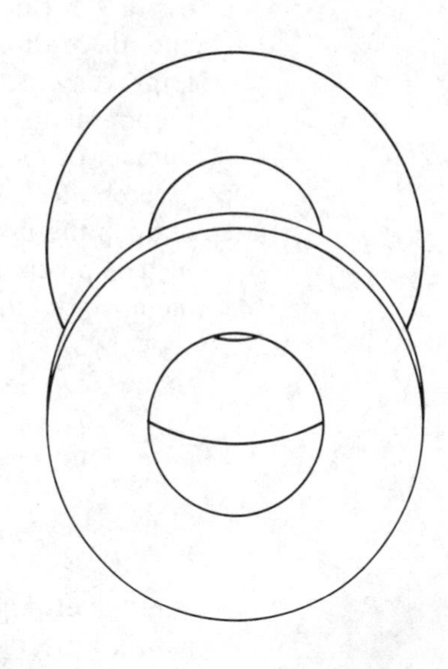

Fig. 6/4 Possible misinterpretation of object and shadows

of binary data are capable of assuming 256 different combinations of 1 and 0. Since one byte of computer memory comprises 8 bits, it follows that any whole number up to 255 can be contained in a single memory location (the 256th combination being all zeros). Thus, if one byte is used per pixel then 256 shades of grey can be identified or, taken one step further, 256 different colours. The principle is illustrated in Fig. 6/5 showing the digitised representation of a simple washer.

Fig. 6/5 Principle of binary image representation using 1 byte per pixel

2	2	1	4	3	3	3	2
2	6	50	76	76	50	4	2
4	50	126	100	100	126	50	5
2	76	100	20	20	100	76	3
4	76	100	20	20	100	76	2
5	50	126	100	100	126	50	1
2	2	50	76	76	50	1	1
1	3	2	4	2	2	2	1

Digitised byte values

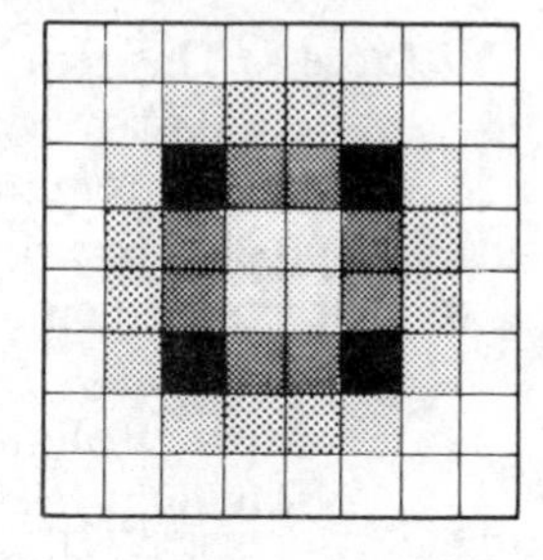

Digitised pixel intensity

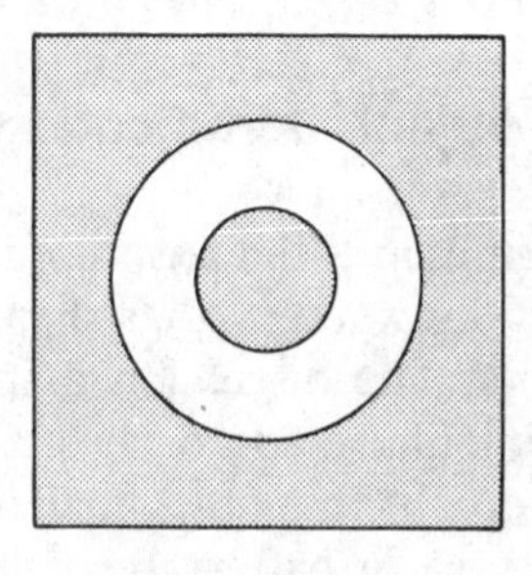

Shape to be digitised

This situation now enables contrast to be identified such that shadows, glare, depth of form, occlusions, etc. can be distinguished from the component form, and three-dimensional imaging is made possible. Although a significant improvement in image capture is recorded by this technique, memory requirements become an important factor. The term **resolution** refers to the number of individual pixels used to form the complete image. Obviously the greater the resolution, the more pixels, and the sharper and more clearly defined the image. A relatively low-resolution image is typically represented by a 64 pixels by 64 pixels matrix array. If one byte per pixel is used then the memory requirements become

$64 \times 64 = 4096$ bytes or 4K of memory

A medium-resolution image, represented by 256 × 256 pixels, requires

$256 \times 256 = 65\,536$ bytes or 64K of memory

A relatively high-resolution image is typically represented by a 512 × 512 pixel matrix array. Memory requirements for this resolution become

$512 \times 512 = 262\,144$ bytes or a massive 256K of memory

These memory requirements are based on the processing of a single image which will be held within an area of memory. If multiple images are required for comparison then the memory requirements grow in direct proportion to the number of images. This is a significant problem since most 8-bit microprocessors have 16 address lines with which to address memory. This gives 2^{16} possible memory locations to which the microprocessor has direct access. This represents 65 536 bytes of memory (64K), clearly inadequate to store a high-resolution image.

Memory requirements may not be the only limitation. If it is assumed that to process one byte of information takes say 1 millisecond (1 thousandth of a second), then typical processing times of a single image will be

Low-resolution image: (64 × 64 pixels) = 4 secs
Medium-resolution image: (256 × 256 pixels) = 66 secs or 1 min 6 secs
High-resolution image: (512 × 512 pixels) = 263 secs or 4 mins 23 secs

For response in real time, these times are unacceptably long and this is why preprocessing of image data is often carried out. Memory requirements and speed limitations can be alleviated if only part of an image is digitised.

6.3/3 Pattern recognition and object identification

Pattern recognition is the term used to describe the process of identifying an object from a knowledge of its shape, make-up and other characteristics and measurements. The discussion that follows is limited to a brief account of certain rudimentary approaches employed in the recognition of simple two-dimensional (silhouette) images.

The first stage in pattern recognition is usually to isolate the foreground image from the background, accomplished by detecting when a significant change in level occurs between adjacent pixels. Since these changes occur at the object's edges, this process is referred to as **edge detection**. At this stage (with gray scale images), insignificant changes in level can be removed or masked out by setting them to the background level, or averaging the levels of a number of pixels using a weighted calculation. There will often be some means to mark (or store) the positions of the edge levels detected.

A second stage may then involve tracking the identified edges by noting the level value of each adjacent pixel that surrounds the one under consideration, and then moving around the image on the basis of following identified foreground pixels. In this way it is possible to identify and form a boundary around the image and thus enclose a silhouette of the object under consideration. From this, it is possible to determine (and further calculate) characteristics such as

a) The greatest 'length', 'width', etc. of the silhouette.
b) The 'perimeter' length of the silhouette.
c) The 'area' of the enclosed silhouette.
d) The 'position' of the centroid of the silhouette.
e) Certain ratios (e.g. area/perimeter, length/width, etc.).
f) The presence of certain unique features (e.g. holes).

These can then be compared against corresponding stored values for the known objects being visualised. In practice a probability level (an accepted level of closeness) is used to indicate a match. This approach works best when the objects being visualised exhibit wide differences in the above characteristics.

The above approach is capable of identifying objects irrespective of their outline shape and it is therefore object identification rather than pattern recognition. Since different components (in size and shape) may exhibit broadly similar characteristics, this approach in itself is not complete enough to differentiate between different shapes. A common method used to recognise the outline shape of an object is **template matching**. This method compares the outline information from the edge processed image with stored representations of known object shapes. The approach is illustrated in Fig. 6/6. (As an exercise, match the object shape shown, with one of the stored representations or 'templates'.) The computer carries out a similar matching process and, for each stored pattern, computes a probability that a match exists. The pattern with the highest probability is then chosen as corresponding to the object being visualised.

Template matching techniques rely on the object being in an identical orientation to that of the pattern with which it is being compared. Since one application of vision systems is to recognise and then bring about orientation of objects, further techniques are required. One technique relies on transforming the position of the digitised image by mathematically rearranging (rotating) the byte patterns. Each new position can then be compared with stored templates, as described above, until a close match occurs. The number of transformations required to achieve the best match is a measure of the rotation the object must undergo to achieve alignment. A second different technique calculates the second moment of area (or second moment of inertia)

Shape to be recognised

Fig. 6/6 Pattern matching by means of stored template data

0	0	1	1	1	0	0	0
0	1	1	1	1	1	0	0
1	1	0	0	0	1	1	0
1	1	0	0	0	1	1	0
1	1	0	0	0	1	1	0
0	1	1	1	1	1	0	0
0	0	1	1	1	0	0	0
0	0	0	0	0	0	0	0

(a)

0	0	1	1	1	0	0	0
0	1	1	1	1	1	1	0
1	1	0	0	0	0	0	0
1	1	0	0	0	0	0	0
1	1	0	0	0	1	1	0
0	1	1	1	1	1	0	0
0	0	1	1	1	1	0	0
0	0	0	0	0	0	0	0

(b)

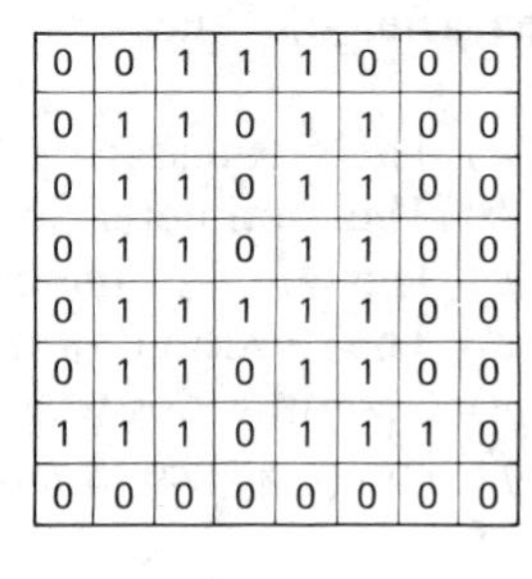

0	0	1	1	1	0	0	0
0	1	1	0	1	1	0	0
0	1	1	0	1	1	0	0
0	1	1	0	1	1	0	0
0	1	1	1	1	1	0	0
0	1	1	0	1	1	0	0
1	1	1	0	1	1	1	0
0	0	0	0	0	0	0	0

(c)

1	1	1	0	0	0	0	0
1	1	0	0	0	0	0	0
1	1	0	0	0	0	0	0
1	1	0	0	0	0	0	0
1	1	0	0	0	0	1	0
1	1	1	1	1	1	1	0
1	1	1	1	1	1	1	0
0	0	0	0	0	0	0	0

(d)

of the object and further refines this with calculations that form a unique 'signature' for the shape. Comparisons with stored values can then be made.

Techniques of vision data analysis are complex and rely on the combination of advanced mathematics and statistics married in many instances to advances in computer hardware and advanced programming techniques. As such robot vision remains an intense area of research and development.

Questions 6

1. Discuss those human senses that it would be desirable to implement within robot systems.
2. Outline the advantages and limitations of the available human senses in the industrial environment.
3. Differentiate between 'internal' and 'external' sensors in the context of industrial robots, and highlight the need for the latter.
4. Discuss the concept of 'levels of variability' in terms of robot senses, and outline the strategies involved in overcoming them.
5. Explain, giving examples, the difference between an 'analog' and a 'digital' electrical signal and state where each type is likely to be encountered.
6. Briefly explain the need for, and the principles of, A-to-D and D-to-A conversion.

7. Define the terms 'transducer' and 'sensor' outlining the principle differences between them.
8. Discuss *five* important characteristics of sensing devices.
9. Briefly discuss the ways in which sensing devices may be classified.
10. Explain the term 'proximity sensing', its implementation and those industrial applications to which it may be applied.
11. Discuss the concept and application of 'range sensing' within an industrial robotic environment.
12. Briefly outline the use of 'strain gauges' in the application of force sensing.
13. Outline the principles of operation of tactile sensing techniques.
14. Suggest likely industrial applications involving the use of acoustic heat and gas sensors.
15. Briefly discuss the concepts of 'polling' and 'interrupts' in the control of external sensory feedback.
16. Explain the relationship between image resolution, memory capacity and processing speed in industrial vision systems.
17. Explain the terms 'pixel', 'byte', 'digitising' and 'preprocessing' and outline their importance in industrial robot vision systems.
18. State *four* essential components of any successful industrial robot vision system.
19. Explain the principles behind image representation using 1-bit per pixel and 1-byte per pixel outlining the relative advantages and limitations.
20. Discuss the principles behind object identification and pattern recognition in the context of industrial robotic applications.

Programming Industrial Robots 7

7.1 Modes of programming

7.1/0 Concept of a stored program

The flexibility of modern industrial robots is derived from the fact that they can be re-configured to carry out a number of quite separate tasks. To achieve this the robot has to be provided with a set of instructions sufficient to enable it to carry out a particular sequence of events. An event may be an axis movement, an instruction to operate an end effector function, a timed dwell, or the analysis of input signals. All computer-based systems operate by carrying out instructions in a strict and ordered manner. A collection of such instructions is termed a *program*. The instructions forming the program must be presented to the robot control unit sequentially in the order in which the programmer requires them to be performed. This is because the control program within the robot control system is designed to execute the program from the first instruction, in strict sequence, to the last instruction. In computer systems the instructions are coded and placed in sequence, starting at a known place within the memory of the control unit. In technical terms, the program is placed at a known *memory address*. The control system program responsible for executing the sequence program will thus always commence its interpretation of the stored instructions from this known memory address.

Since the instructions are retained within computer memory they are known collectively as a **stored program**. Stored programs are often termed *software*. A prime advantage of this arrangement is that the contents of this memory (the program) can be copied (SAVEd) to a backing store, such as a magnetic tape or disc. (The term software is often extended to include the media on which programs are stored.) This enables the same program to be re-copied (LOADed) back into memory for use at a later date. The process of generating a program is thus an investment in time and effort that can be realised over and over again. A second advantage is that facilities are normally available within the control system software to edit the sequence program resident within computer memory. Thus, sequence modification and fine tuning of the robot's actions can be carried out quickly and conveniently at the workplace. The edited version of the sequence program may then be saved to backing store as current.

Industrial robots are unique in that they can be programmed in a variety of ways. More importantly perhaps, they may be programmed by people with little or no knowledge about computers or programming. This may be essential where robots are to be taught tasks previously carried out by skilled operatives, such as spray painting. The skilled sprayer, the obvious choice for teaching the technique and dexterity of the task, may not know (or even want to know) anything about the intricacies of the computer control system and how it should be programmed. Because of the availability of backing store devices, once learned the task does not have to be re-programmed or re-learnt after long periods of no use. Human operators would be likely to require a re-learning period after a long lay off from a task, and there is no certainty that a particular operator could in fact learn a particular task.

7.1/1 Programmable control

A program may be defined as a collection of instructions arranged in such a way as to achieve, in the case of a robot, controlled movement through a desired sequence. In an industrial robot the desired sequence is likely to be a repetitive task which performs work useful to the organisation. When the sequence of instructions, and hence the controlled movement, can easily be changed or modified it is termed **programmable control**, and the robots are said to be re-programmable. Early robots were 'programmed' to perform strict sequences according to mechanically operated limit switches and end stops which had to be manually positioned between the maximum and minimum limits of travel of the various robot axes. No feedback of position or velocity was involved and conditional actions (the ability to perform slightly different actions depending on different conditions) were not available. These robots were often termed non-servo robots, sequence controlled robots or bang-bang robots since they would physically bang against limit switches at the limits of their movement. Although the sequence could be altered by re-positioning the limit switches and end stops, it is debatable whether these robots can be called truly programmable.

Modern industrial robots controlled by digital computer based control systems offer true programmability under software control. Software control means that the instruction sequences and commands are resident in computer memory whilst the task is being executed. There is no reliance on external switching elements, or stops, to determine the action of the robot, although they may be used as feedback devices or sensors. Modern industrial robots are essentially playback machines. The sequence program instructions may be generated, or originated, in a number of ways. Programming can be carried out at the workplace, involving the robot, or remote from the workplace via a computer terminal and specialist robot programming languages. The former is known as **on-line programming** and the latter as **off-line programming**. Within these techniques four common modes of programming industrial robots can be identified. Their application, relative advantages and limitations are discussed in the following sections. It is argued that the choice of an industrial robot should *not* be based on programming capability alone since the amount of time spent programming the robot is disproportionately low in comparison to the amount of time spent performing its allotted task.

7.1/2 Lead-through programming

Lead-through programming is an on-line programming technique. It is the oldest form of programming applied to industrial robots. The programmer, usually a skilled operative, physically leads the robot through the desired motion sequence. In so doing the operator is 'teaching' the robot the correct sequence of moves for each of its axes. It is sometimes termed *lead-by-the-nose programming*. A special detachable handgrip is usually fixed to the end effector of the robot to facilitate programming. In the case of heavy robots, counterbalancing may be required to enable the operator to achieve smooth movements. Hydraulically operated robots may similarly require pressure relief.

This technique is usually employed (perhaps incorrectly) on the basis of the following assumptions:

a) The skilled operator, knowledgeable in the task, can impart his or her skills, dexterity and technique to the robot.
b) The senses of 'feel' of the operator can be passed to the robot.
c) The operator, expert in the task, is the best teacher.
d) The method most favoured by the human operator is the most applicable method for robot operation.

In most cases the robot arm is usually more heavy and cumbersome to manipulate than for example a tool or spray gun. Thus, the dexterity, feel, speed and flowing movement of the operator will almost certainly be severely impeded and therefore nullify many of the above assumptions. In many cases this technique is inappropriate because of the size and weight of the robot. In addition, lead-through programming cannot be used in hazardous environments, since adequate protection of the operator is probably impossible. Synchronisation with other operations within the workplace is often difficult to achieve since all the actions are determined by the operator. Programming can only be accomplished at 'human speed'. Some control systems may have the facility to playback the program sequence at different rates of execution.

Lead-through programming has the advantage that robot motion is directly determined, under the control of the operator, within the three-dimensional workplace within which it operates. Complex mathematical descriptions of arm movement are thus eliminated; inadvertent collisions and unexpected departures from intended movements are eliminated; and any inherent inaccuracy of the robot control system can automatically be compensated for by the operator. The actions of the robot under this mode of programming effectively mimic the behaviour of the human operator. There is little need, or opportunity, for real time intervention by external sensors. This factor alone perhaps points to those applications where this technique of programming is applicable.

Actual programming takes place by the control system periodically reading (termed **sampling**) and storing the status condition of each axis. Positional and/or velocity data for each axis are recorded and stored in computer memory. Subsequent readings are regularly obtained and stored sequentially in memory, thus building up a program of positional and velocity data corresponding to the movement sequence being taught. The control system records the coordinate data in one of two ways. It is possible to read and store the information at *the end of each discrete movement*. The end position of each movement will be signalled by the operator normally by pressing a 'teach' control. This technique is suitable for relatively simple point-to-point motions. Alternatively, the control system can successively record data at *fixed time intervals* throughout the performance of the task. This is known as programming in real time and will be employed where contours have to be traced or where some measure of dexterity, feel or close control has to be taught.

Lead-through programming is relatively inefficient in terms of control system and memory utilisation. For example, the length of the program (and the amount of memory required to store it) is directly proportional to the time required to complete the task. There are no facilities for repeating commonly used actions. The same actions are recorded, in full, each time they are executed. This occurs even though an identical sequence of instructions may already be resident in memory through an earlier recording of an identical sequence. When programming by this technique the robot is effectively rendered non-productive since it cannot be performing a worthwhile task at the same time that it is being programmed. The concept of lead-through programming is illustrated in Fig. 7/1.

7.1/3 Drive-through programming

Drive-through programming is also an on-line programming technique. As with lead-through programming it requires the use of the robot during program generation. It is also sometimes called *walk-through programming*, a term not generally preferred since it can easily be confused with the previous programming method. The technique requires the operator to drive the robot through the desired movements by remote control. It is thus suitable for programming even the largest, heaviest robots. The operator is usually in close proximity to the robot, effecting control via a hand-held **teach pendant**. The technique is perhaps the least safe of all programming methods. A potential safety hazard is created by the operator being in reasonably close proximity to the robot. There is the obvious likelihood of the operator pressing an incorrect button on the teach pendant causing an erratic or unintended movement. To minimise such eventualities, programming is often carried out under reduced or restricted speed control. The teach pendant will have a range of controls and functions relating to: axis movements; end effector control; programming and execution controls; and various status display indicators. A typical teach pendant is illustrated in Fig. 7/2.

Although this method eliminates the need to physically haul a heavy robot arm around the workplace, there is a distinct lack of fine control available due to the lack of feel and the comparative remoteness of the operator. Additionally, it is difficult to precisely predict the target position of the end of the arm. Absolute accuracy of robot movement is to some extent of secondary importance since it can be compensated for by the operator during programming. Axes may be driven either independently or simultaneously throughout the operation sequence as desired. Since the operator can be positioned remotely from the robot, this technique can be applied in those environments considered hazardous to the operator. A disadvantage of using

Fig. 7/1 Lead-through and drive-through programming techniques

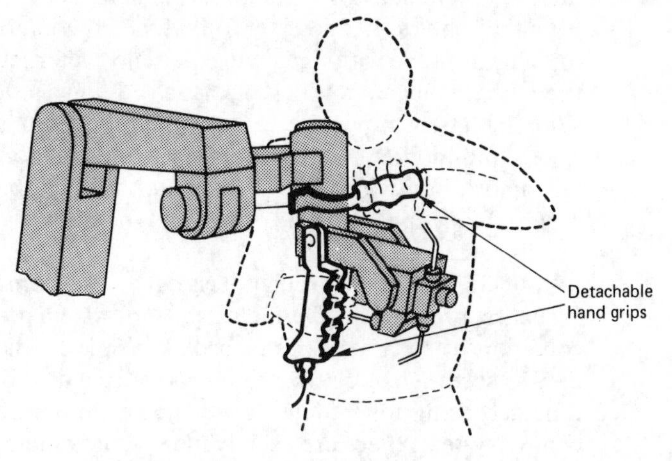

(a) **Lead-through programming**

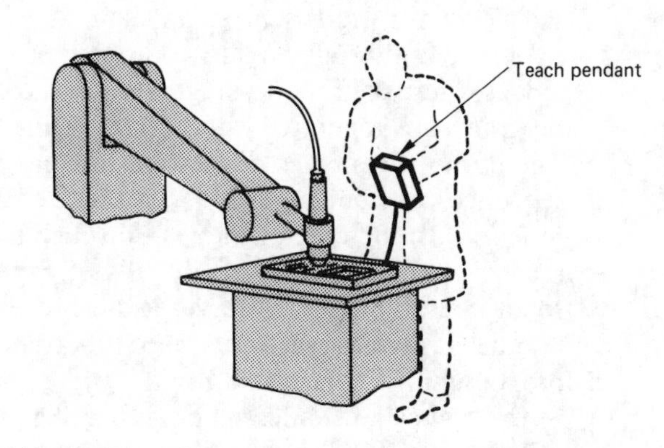

(b) **Drive-through programming**

Fig. 7/2 A typical teach pendant

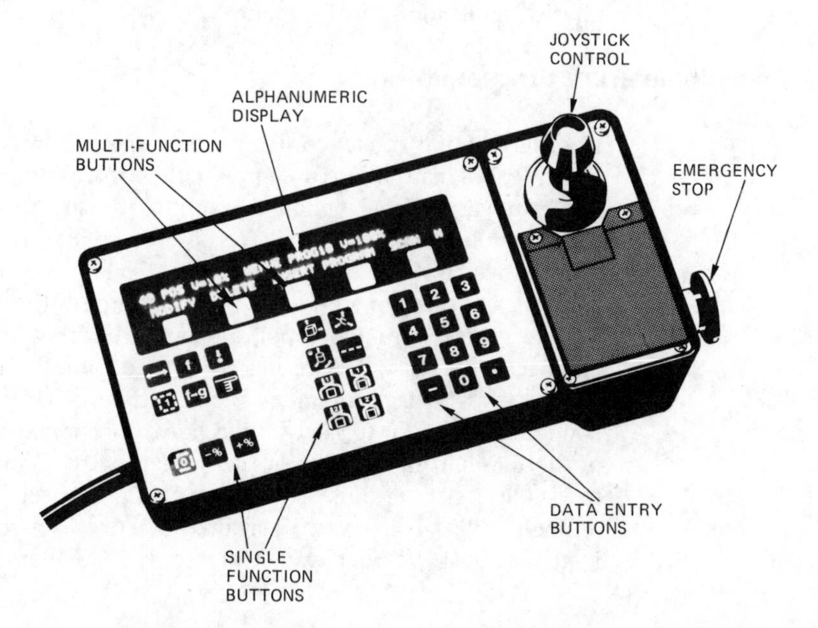

the teach pendant is that the programmer has to continually divert attention from the robot to the pendant to find the function buttons to press. Although this may improve with time and experience, alternative approaches have been tried. These include joysticks and a scaled-down model simulator of the robot. With the latter, moving the axes of the simulator causes the axes of the robot to respond in like manner. In both cases the actions of controlling the robot are almost intuitive and they eliminate the need to search the keypad for the required functions. Both joysticks and simulator models are light and easy to manipulate.

As with the lead-through technique, the control system generates the sequence program by periodically, and systematically, sampling the status condition of each axis and recording positional data. Similarly, coordinates may be recorded at discrete intervals (as signalled by the operator), or in real time according to predetermined time sample intervals. Depending on the control system, there are usually more programming features available to the programmer via the facilities of the teach pendant. For example, there may be facilities to jump to different parts of a program (either conditionally or unconditionally) to repeat existing move sequences or actions, facilities to invoke timed delays and execute often-used subprograms, to read input data signals from various sensors and so on.

Both lead-through and drive-through programming are essentially teaching modes where the robot is taught its actions by the operator. As such, sequence programs are generated rather than originated by conscious program writing. The prime advantages of these techniques are the ease with which program sequences can be originated and the fact that skilled programmers are not required. In the case of the lead-through method it is valuable to be able to readily apply the time-learnt skills of experienced practitioners directly to robot operation. It also serves as a good example of re-deploying operators to other tasks as a consequence of robotisation. The concept of drive-through programming is illustrated in Fig. 7/1.

The following modes of programming concern techniques of originating sequence programs by pre-conception rather than having them generated by physical movement of the robot.

7.1/4 Coordinate entry programming

A programming mode similar to drive-through programming is **coordinate entry**. Robot instructions are input to the control system, as lines of a sequence program, via the teach pendant. Positional data relating to particular axes, and other command instructions, are entered sequentially and discretely in the order in which they are required. However, no movement of the robot takes place during this program entry phase. Some controllers allow program entry to take place whilst the robot is busy performing another task. Coordinate entry may thus also be considered as an off-line programming technique. A difficulty with this technique is that human programmers often tend to be happier thinking in terms of 'tidy' linear or circular movements, or in terms of the familiar movements of the human body. Whilst it is often quite easy to visualise a three-dimensional movement by itself, it is considerably more difficult to visualise how to coordinate 3, 4 or 5 axes simultaneously to achieve it in the most efficient way. Conversely, a movement seemingly requiring multiple-axis movements may be accomplished by movement of a single axis. For example, for humans to transfer an object standing on a table in front of them to a similar position on a table behind them may involve a number of simultaneous body movements. One can imagine the wrist, elbow, shoulder and waist all being moved together. For a revolute arm robot however, one simple sweep of the shoulder ('over its head'), through 180°, would probably accomplish the same task. Simple movements, whilst being easy to visualise and describe by human programmers, are not entirely conducive to efficient operation. Programmed moves will take longer to execute than they should – there will be redundant motions and so on.

Many control systems allow entry of positional data to be made in rectilinear and polar coordinates but they have to be mathematically transformed, by the control system software, before achieving the desired motion. The term **transformation** refers to the process of converting positional data, as understood and supplied by the programmer, into the data required to be fed to each of the robot axis manipulators in order to achieve the desired position and orientation of the end effector. Additionally, it allows the position and orientation of the end effector to be determined by the relative positions of the robot joints. It follows that some previously (agreed upon) defined datum or zero position for each axis and the end effector must be used as a basis from which descriptions are made. Transformation is one function of the master operating system software within the robot control unit.

Coordinate entry is inherently safe since no movement of the robot need take place during program entry. The operator can retire to a safe distance before initiating a cycle start to execute the stored program. Unless highly organised, this method is a somewhat tedious, almost trial-and-error method of programming and as such is relatively inefficient. The technique does have other wider applications than purely as a programming mechanism. It is a technique that may be employed for editing, entering conditional moves, programming interlocks, actuating end effector functions, etc., during other programming modes and during program proving operations. The concept of coordinate entry programming is illustrated in Fig. 7/3.

7.1/5 Remote off-line programming

Remote off-line programming refers to programming carried out completely away from the robot. The sequence program is written, purged of errors (de-bugged) and proven before it is loaded into the control system memory of the robot. Because programming is done in advance, and not during teaching, it may also be termed **pre-programming**.

Off-line programming is made possible by the use of specialist robot programming languages. Since such languages are usually written for specific applications or for certain robot families, there is no absolute standard for a robot programming language. Off-line programming is the most recent programming technique to be developed for industrial robots. It seeks to overcome many of the limitations of the previous methods by allowing high accuracy, synchronisation with other elements within the production system, and appropriate reactions and response to sensory feedback. The claimed advantages of off-line programming techniques include the following:

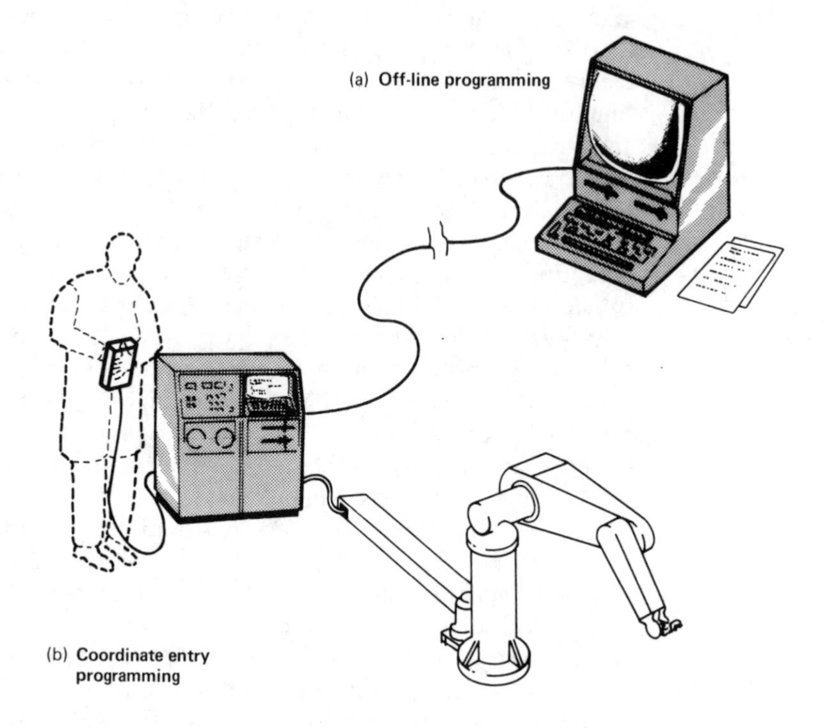

Fig. 7/3 Coordinate entry and remote off-line programming techniques

a) Programs can be prepared without the need of a robot.
b) Robot operation, synchronisation and movement sequences can be carefully pre-planned and optimised.
c) Programs can be written in a modular fashion allowing library routines to be built up for use in other programs.
d) Programs can be proven in advance of being run by the robot (via *computer simulations*), lessening the risk of damage and physical injury.
e) Conditional operation of the robot can easily be built into the program dependent on the feedback of sensory devices, thus creating an *adaptive system*.
f) Existing CAD/CAM information can be more easily incorporated into the control functions available.
g) They may have facilities for controlling other automated elements within the workplace.
h) English-like programming languages allow quick uptake of programming skills and simplify program production, editing and de-bugging.

Some robot programming languages have been developed as completely new, purpose-designed languages. These may have been specifically developed by robot designers and manufacturers for a family of robots, or by research establishments, to include a range of specific data processing and control capabilities. Other languages have been developed by modifying and/or extending existing computer programming languages. This approach has the advantage that existing implementations of the language will be readily available, proven, tested and available to run on a wide range of inexpensive computer systems. Although completely new command words and programming structures will be available, some compromise is inevitable in the speed, efficiency and facilities that can be implemented. Yet other robot programming languages are in fact programs themselves written in existing computer programming languages. This is the simplest, least costly and quickest approach to develop. The robot language will consist of written subroutines which are called in response to certain key words. The obvious limitation is that the robot language will be restricted to those features implemented in the parent language.

Off-line programming may take two distinct forms. The first, similar to the coordinate entry approach, requires **target positions** for the robot moves to be identified in terms of X, Y and Z coordinates. This has the disadvantage that all the coordinates to which the robot must move, must be known. This may require several measurements at the workplace to ascertain the various target positions. Once known, however, the sequence of motions can be simply repeated at other positions within the workplace by shifting the datum start position. Similar, consecutive or 'handed' operations can thus be performed for very little extra effort. The second approach is known as **world modelling**. In this technique the robot is first taught the positions of all the important points within its workplace. Similarly it is taught the shapes, gripping points and orientations of all the components it is likely to handle during the course of carrying out its programmed tasks. These positions, components and certain often-used collections of movements are all identified by a certain keyword or phrase. Programming consists of sequencing these phrases in a logical, English-like way thus enabling programming to be learnt and accomplished quickly. For example, programming instructions may take the form of

APPROACH CONVEYOR
PICK LARGE BRACKET **FROM** CONVEYOR
APPROACH JIG
PLACE LARGE BRACKET **IN** JIG
APPROACH FEEDER
PICK SMALL BRACKET **FROM** FEEDER
APPROACH JIG
SET SMALL BRACKET **TO** LARGE BRACKET and so on.

The concept of remote off-line programming is illustrated in Fig. 7/3.

The coordinate system used for a world modelling approach tend to be conventional X, Y and Z coordinates orientated about the major axis of the robot. They may also be termed *world coordinates*. A useful extension, available in many control systems, allows a similar X, Y and Z coordinate system to be defined about the major axis of the end effector. This enables the complete X, Y, Z coordinate system to become re-orientated about the tool or end effector. It is then possible to switch between systems to enable dimensional moves of the robot to be made relative to the tool or end effector rather than the original datum. To differentiate between the two coordinate systems, the latter are termed *tool coordinates*. Only one system can be operative at one time. World and tool coordinates are illustrated in Fig. 7/4.

Offline programming techniques cannot know, and hence make any compensation for, the weight of the end effector and/or cargo.

Fig. 7/4 World and tool coordinates

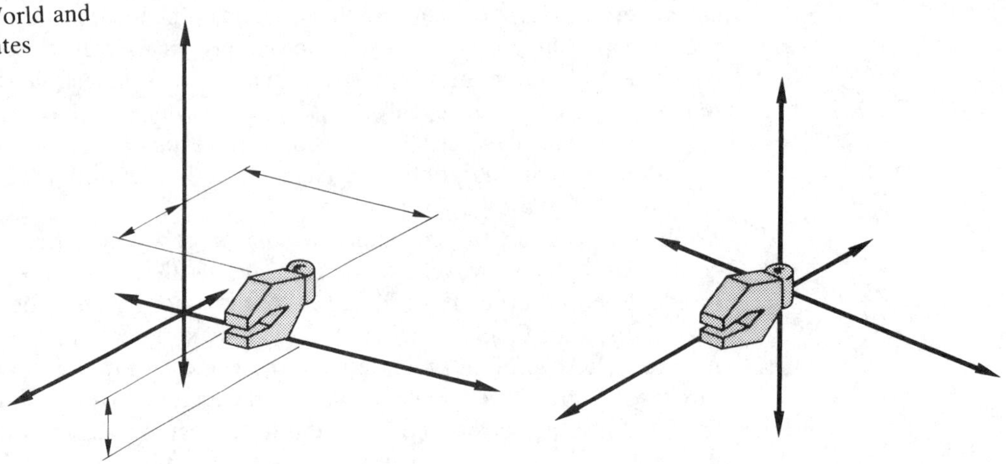

(a) WORLD COORDINATES: absolute axis zeros; all movements referenced from this datum plane

(b) TOOL COORDINATES: datum references made coincident with tool/end effector attitude; all movements referenced from this reference plane

7.2 Programming facilities

7.2/0 Control system facilities

Most control systems comprise a number of operating modes that can be switched in and out by the operator. If a VDU display is available within the control console, various items of information will be displayed in each mode. Although different control systems offer different facilities a typical range will include the following modes:

HOME mode This mode enables the robot axes to be driven, either independently or together, to their datum or home positions. However, because a return-to-home position may be taken by the shortest route, there is danger of collision of objects within the working envelope, and it is recommended that before a 'return to HOME' command is issued, the robot should be jogged under manual control to a position close to its home position. Home searches may also be accomplished in a single direction (either positive axis motion or negative axis motion) only. Thus, the robot axes may need to be jogged to a position which can accommodate this. The VDU display in HOME mode is likely to include such information as:

a) Joint positions
b) Coordinate positions
c) Present datum positions
d) The values of inbuilt timers and counters
e) The status of the input/output bits . . . and so on.

TEACH mode This mode enables the programmer to enter a sequence program, edit a sequence program or save a sequence program held in memory. Program entry may be made: by coordinate entry via the control console or teach pendant; via one of the dynamic modes of programming with positional data being periodically sampled and stored; or by loading in from a backing store device. There will be facility within the teach mode to manipulate the robot axes to any convenient position under manual control, set the values of inbuilt timers and counters, and enable or disable input/output bits. Screen display information includes such information as:

a) Sequence program data currently held in memory
b) Current joint positions
c) Current coordinate positions
d) Monitoring of inbuilt timer and counter values
e) Monitoring and status of input/output bits . . . and so on.

STEP mode This mode enables the stored sequence program to be executed one program step at a time. Subsequent steps will only be executed when a button on the operator's console or teach pendant is depressed. Screen information is likely to include:

a) Step number being executed
b) A number of program steps before and after the current step
c) Current coordinate position
d) Current joint positions . . . and so on.

AUTO mode This mode enables automatic execution of the stored sequence program. Most of the comments for STEP mode apply equally well to this mode. The screen display information is also likely to indicate whether or not the automatic program repeat function is operative.

PARAMETER mode This mode enables a number of internal settings, status values and default values to be defined and determined. Many inbuilt parameters may concern the setting and technical operation of the various system components, such as the servo drives for example. These parameters are set by the manufacturer and should not be altered. Other parameters may be user-definable and set to suit the requirements of the end user. Typical parameter settings include:

a) The maximum limits of travel for each axis, after which motion ceases and an alarm condition indicated.
b) The maximum operating speeds for each axis in both single step and automatic execution modes.
c) The datum or home position for each axis.
d) End effector offset values relative to the end effector mounting arrangement.
e) Transmit and receive setting for the RS232 communications interface.

Some control systems may restrict entry into some modes until a 'return to home' command has been issued. This prevents functions such as TEACH,

STEP and AUTO being activated from unknown axis positions, and also the resumption of sequence program execution from unknown axis positions after an emergency stop has been carried out. Control system facilities by mode of operation are illustrated in Fig. 7/5.

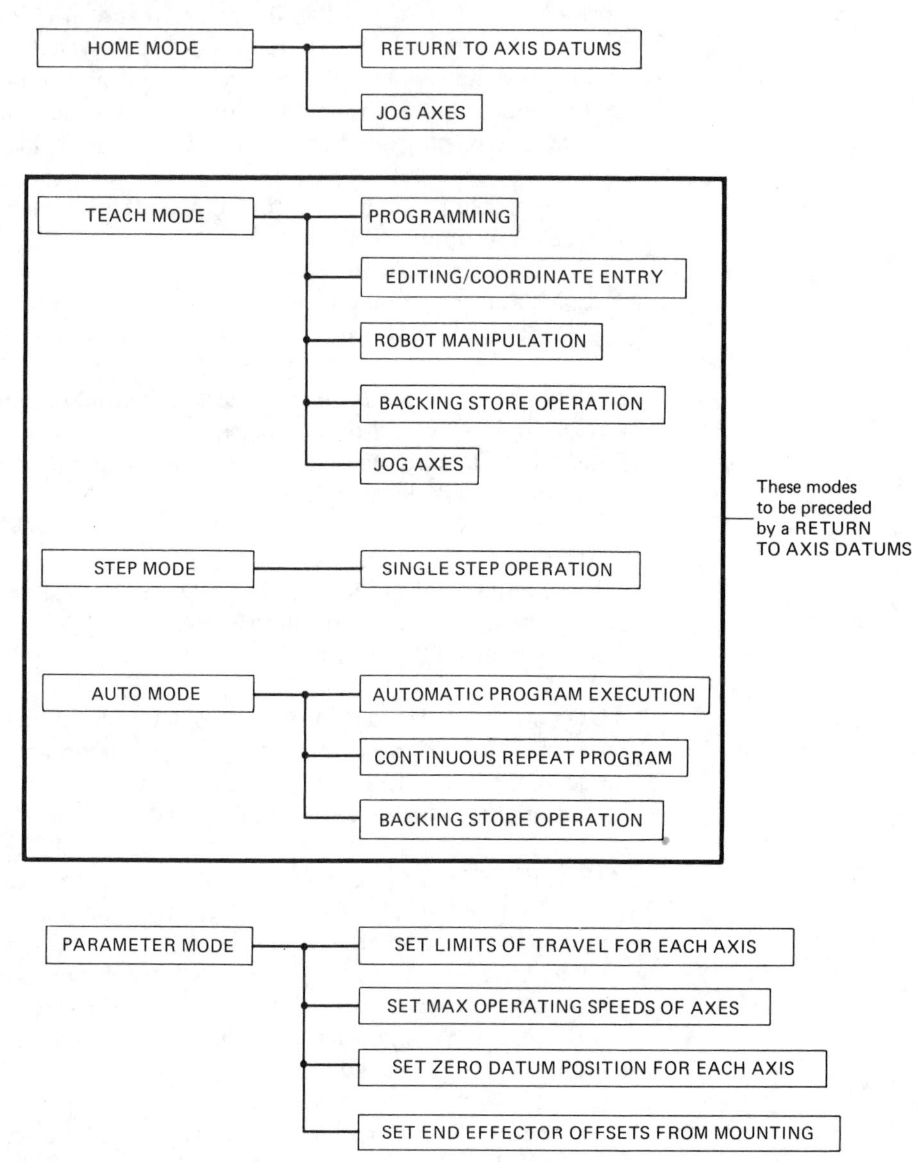

Fig. 7/5 Robot control system facilities by mode

7.2/1 Robot startup procedures

Although different industrial robots have different facilities, functions, modes of operation and control, it is a wise precaution to adopt a standard procedure of safety checks before initiating the robot to operate. Prior to robot startup the following points, as a minimum, should be observed:

1 Remove all unwanted items from within the robot working environment.
2 Ensure that the end effector is not holding or gripping a component.
3 Take note of the position of all jigs, fixtures and other likely obstructions within the robot workspace.
4 Ensure that the teach pendant or robot controller is positioned such that the operator, when using the controller, is outside the working envelope of the robot.
5 Ensure that no other person is inside the safe zone (designated a safe distance outside the working envelope) and that all safety interlocks are correctly armed.
6 Check that all connections to the robot are secure and that there is no damage to cables and service pipes.
7 Check that the robot mountings are safe and secure.
8 Check that hydraulic/pneumatic pressure to the robot is correctly set and that couplings and joints are leak free.
9 Check that fluid levels are correct where appropriate.
10 Check that trailing leads, hoses and cables do not present a hazard.

Once the robot has been safely started it should be jogged to a position near to its home position and a 'return to HOME' command issued. This ensures that the robot arm elements are in a known position with which the operator is familiar, and also that the control system has a known datum point about which all axis movements can be referenced.

Programming industrial robots from the point at which the robot has been started up and is at rest in its home position is likely to be individual to a particular industrial robot and the specific operating and programming manual for the robot concerned should be consulted. A typical sequence of events would be:

a) Erase the data held within the control system memory.
b) Set various parameters for the task to be programmed, for example
(i) maximum axis speed values
(ii) limits of axis travel
(iii) axis zero datums
(iv) tool or end effector dimensions from the mounting plate
c) Move the robot manipulator to the first location using the teach pendant and store axis displacement values.
d) Program in any required conditional signal responses.
e) Program end effector activation/manipulation movements as required.
f) Repeat steps *c* to *e* until program sequence is completed.
g) Return to home or start position under manual control.
h) Check all location, sequence, input and output data in memory.
i) Re-check parameter values and settings.
j) Carry out startup checks (points 1 to 4 above).
k) Run under single step mode.
l) Run under single cycle auto mode.
m) Run under continuous cycle auto mode.

7.2/2 Levels of software

In introducing the concept of a stored program mention has already been made of two types of software. First, the user-defined sequence of instructions that causes the robot to execute a given task, we have called the **sequence program**. The sequence program is essentially data arranged in a specific format. Secondly, the inbuilt control system program that actually executes the instructions of the user's sequence program, we have called the **control program**. The two types of software are distinctly different and should not be confused. For example, the control program is part of the *system software* designed by the control system designers. It addresses the microprocessor itself and is written in machine code. It cannot be altered by the user and as such is stored in non-volatile, read only memory (ROM). The system software contains other sub-programs for carrying out other specialist tasks required for the overall operation of the robot. Such tasks include:

a) Accepting, decoding and storing command, function and positional data from a number of sources.
b) Performing calculations and coordinating axis movements.
c) Providing editing facilities for the sequence program.
d) Providing communication information via display updating, etc.
e) Accepting, decoding, coordinating and processing feedback signals from on-board and external transducers.
f) Coordinating and distributing output signals to manipulator and end effector elements.

Collectively this system software is known as the **operating system**. The operating system is an example of *low-level software* since it is written in a language that can be directly understood by the microprocessor in the control system.

Programming the microprocessor directly represents the lowest level of programming possible, and for this reason the language is commonly termed *machine language* or **machine code**. Machine code comprises binary coded bit patterns arranged to form commands that the microprocessor can understand, or data. Programming in machine code is the most versatile form of programming since it enables all the facilities of the microprocessor to be fully exploited. Often, when programming complex systems (such as a complete operating system for a robot controller), it is inconvenient to use the low-level language of the microprocessor. It is tedious and time consuming; it requires specialist knowledge; it is often cumbersome to develop and test; and since long binary sequences of ones and zeros have to be encoded, the system is prone to errors. In most cases English-like languages have been developed to overcome these deficiencies.

The second level of software available for programming is called **assembly language**. In this technique, each binary sequence representing a microprocessor instruction is given its own three-character description. For example, JMP is the description given to the instruction to JuMP to another part of the program, CMP is the description given to CoMPare the contents of a single memory location with either a data value or the contents of another memory location, and so on. Each three-character descriptor is formed from the letters of the instruction to which it relates. Thus, assembly language instructions are called *mnemonics*. Programming in assembly language is easier and quicker to achieve than in pure machine code, and the resulting program is easier to read and understand. Once the code has been finalised, it is interpreted by a program called an *assembler* (from which the name of the language is derived), converted into machine code binary patterns, and stored in the correct memory locations ready for execution. Although much easier to work with than machine code, assembly language still tends to be a little tedious to program and difficult to de-bug. A third tier of programming languages has developed which uses more meaningful words, clauses and phrases to describe the actions required.

They are called **high-level languages** since they are relatively quick and easy for (higher level) human operators to understand but impossible to feed to the microprocessor itself without first being *interpreted* in some way. This interpretation process can incur a slight time penalty during execution. In addition, the facilities available to the programmer are limited to those facilities within the particular high-level language. Facilities have to be limited since significantly more memory is required to hold the vocabulary and routines of the language. These are significant reasons why they are not usually used to program and develop control system software. They form a sort of half-way house between being easy and quick to program and develop whilst not sacrificing too much in terms of speed of execution, storage space and available facilities. The implementation of a high-level language usually involves a set of defined keywords, with or without additional supporting information (arguments), invoking a number of machine code sub-programs to carry out the operation suggested by the keyword description. Thus, one keyword in a high-level language expresses the information required to carry out a large number of machine code instructions, hence simplifying the process of programming. Whilst high-level languages are not usually used to develop control system software, they form the basis of nearly all computer languages, and many robot specific programming languages. Their advantage lies in the relative ease and speed with which operators can learn the languages and the ease and speed with which sequence programs can be developed, tested, modified and edited. A number of high-level languages have been developed, each with its own characteristics and each to exploit certain features required in certain applications. Their vocabulary and syntax are all different and each reflects those facilities required in the applications for which they were designed. For example, they include:

BASIC	(Beginners All-purpose Symbolic Instruction Code)
FORTRAN	(FORmula TRANslation language)
PASCAL	(named after the French mathematician Blaise Pascal)
LISP	(a LISt Processing language)
COBOL	(COmmon Business Oriented Language)
ALGOL	(an ALGOrithmic Language)

7.2/3 Robot programming languages

Robot programming languages are examples of high-level languages specifically designed for operating and controlling robots. They allow the relatively easy development, testing, modifying and editing of sequence programs, off-line, usually via stand-alone microcomputers. Because they have been designed specifically for robot applications, their vocabulary reflects those movements, facilities and operations which robots are commonly called upon to perform – for example, MOVE, PICK UP, WAIT and so on. Whilst some robot programming languages have been specifically purpose designed, many have been formed by modifying and augmenting existing high-level languages.

There is a wide variety of robot programming languages currently available. Because of their diversity and their manufacturer or machine-specific nature, only those features and facilities common to almost all programming languages will be described here.

In all high-level languages, in addition to the particular vocabulary, three basic programming structures are supported. It is these structures that give the high-level languages their power and flexibility. Although the keywords that implement the structures vary from language to language, a knowledge of their purpose should enable a better understanding of the basis on which programs are constructed, and offer a smooth transition from learning one language to another. The programming structures are as follows.

1 SEQUENCE
The concept of sequential execution or program instructions underlies most programming languages. Unless specified otherwise, the instructions that form the program will be executed one at a time and in the order in which they are presented. To formalise the concept some high-level languages require that each program line be identified by line or sequence number. The line number indicates the relative position of the instructions in the program sequence, and indicates the order in which the instructions are to be executed.

2 REPETITION
Many highly defined sequences, to which all computer-based equipment is ideally suited, are based on the repetition of a small number of operations. There is a subtle difference between doing a large number of operations and doing one operation a large number of times. For example, the robot control system itself, when running a sequence program, does nothing more than repeat a small number of steps a large number of times. They can be simply illustrated as follows:

1. Get a line of the sequence program.
2. Execute the line.
3. Is this the last line of the sequence program?
   If not return to step 1.
4. Return control to the operators console.

The above sequence will be repeated as long as there are instructions in the sequence program to be executed. Implicit in this simplified example is the operation of the first structure, i.e. sequence. Step 1 must be carried out before step 2, step 2 before step 3 and so on.

3 BRANCHING

The undoubted power, and the often mistaken 'intelligence', of computer execution is derived from the ability of a program to branch (or jump) to another part of the program to continue execution. In the above example, we can see the branch concept in operation. At step 3 a decision is taken as to whether the program line just executed is the last in the sequence program. If it is not the last line, there must still be lines to execute and so the program branches to step 1 to continue execution. Branches may be either backwards (as shown above) or forwards to a point later in the program. The above branching situation is known as a **conditional branch** since the decision to branch depends on the outcome of a particular condition. The condition above is that the line just executed is the last in the sequence. There are two important points to note regarding conditional branches. Firstly, the condition in question can have only two outcomes: yes or no, pass or fail, true or false. This is consistent with the two-state working of digital (binary) operation. Secondly, if the condition is not true, and the branch does not occur, then the program will carry on, in sequence, to the next step – in the above example, by returning control back to the operator. Conditional branches may also be thought of, although not strictly true, as being the decision-making capability of the language.

Most languages support a second branching structure called an **unconditional branch**. When an unconditional branch instruction is encountered (commonly called a Jump instruction), then the program will jump immediately to the part of the program specified. Within a sequence program, branches may be made to a specific sequence line number or to a user-defined meaningful reference word called a *label*, depending on the programming software employed.

There may also be facility to make use of inbuilt *counters* that can be made to increment or decrement under certain conditions and also to set, reset and branch on the values of internal counters to effect timed delays for example.

4 Within the above programming structures the language will also support different **data types**. The use of different data types allows the programmer a measure of planning, flexibility and error reduction to aid program production and development. Common data types are INTEGER, REAL NUMBER, CHARACTER and LOGICAL. Integer numbers are whole numbers only and any decimal portion is simply ignored. The advantages of using integers are that programming can be simplified by not having to build in routines to deal with an unwanted decimal portion, execution is often quicker than with decimal numbers, and there is some saving in memory storage space. Real numbers are numbers that carry a decimal portion. Both integer and real numbers can have arithmetic operations carried out on them. The Character data type allows any valid character (including digits) to be used. Care must be taken however, since numbers stored as character data do not possess any numerical significance and cannot be processed arithmetically. The facility is allowed since there are many occasions when identifying information is a combination of alpha and numeric characters — for example, order numbers, batch numbers, component identification numbers and so on. Logical data types comprise values that are either True or False, often denoted simply by the letters T and F respectively. Their value lies in the ability to construct complex Boolean logical expressions. (See Chapter 4 for further discussion on logic and logic operations.)

7.2/4 Sequence program development

The task to be performed by the robot will necessarily determine, to a large extent, the sequence of events to be programmed. The development of sequence programs however, especially those of considerable complexity, can be significantly aided by a systematic approach. Sequence programs conceived and written at the keyboard will be less efficient and less understandable than those whose logic has been thoroughly thought through.

A planning technique known as **flowcharting** helps to concentrate the programmers thoughts on achieving an efficient operation sequence in a relatively short time. It offers the following benefits:

a) It concentrates the mind onto the task to be performed.
b) It fosters the production of a planned logical operation sequence and highlights areas of duplication, flaws in logic and poor program flow.
c) It can be used as clear, concise and meaningful documentation for the task being developed and for permanent record keeping purposes.
d) It can be used as a medium of communication to those who need to know what the robot is doing, but may not be familiar with the robot programming technique itself, for example customers, managers and operators at shift changeovers.

Flowcharts are diagrammatic representations, on paper, of the logic to be followed by the sequence program. They are constructed using certain symbols joined together by straight flowlines. The flowlines have arrows marked on them to indicate the logic flow. A small number of common flowchart symbols, adequate for most sequence descriptions, are shown in Fig. 7/6. To illustrate the application consider that ten small parts have to be lifted from a conveyor and placed on a tray in a 5 × 2 rectangular placement pattern. A flowchart illustrating the logic of a sequence program to perform this task is shown in Fig. 7/7. It is left as an exercise to identify the programming structures of sequence, repetition and branching which are evident within this simple flowchart, and to comment on whether the sequence can be improved. The important advantage of using flowcharts is that they are language independent. They convey thought processes and logic which are transferable between any programming language or robot control system. It should be a relatively simple process to convert the flowchart into a programmed sequence. Flowcharting is only one of a number of such techniques designed to aid the construction of correct and reliable programs.

Once the logic of the sequence has been ascertained, it has to be converted into a valid and correct sequence program that will run without error. For all but the simplest of tasks a programming concept called **modular programming** can greatly speed up program development. Modular programming advocates that the whole task sequence should be split down into discrete, easily

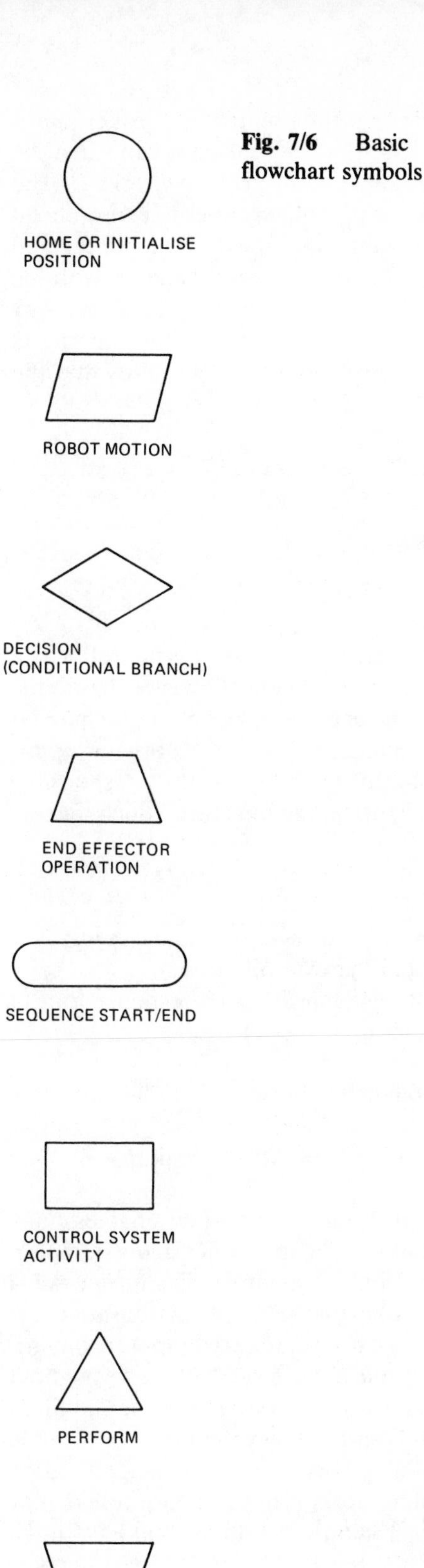

Fig. 7/6 Basic flowchart symbols

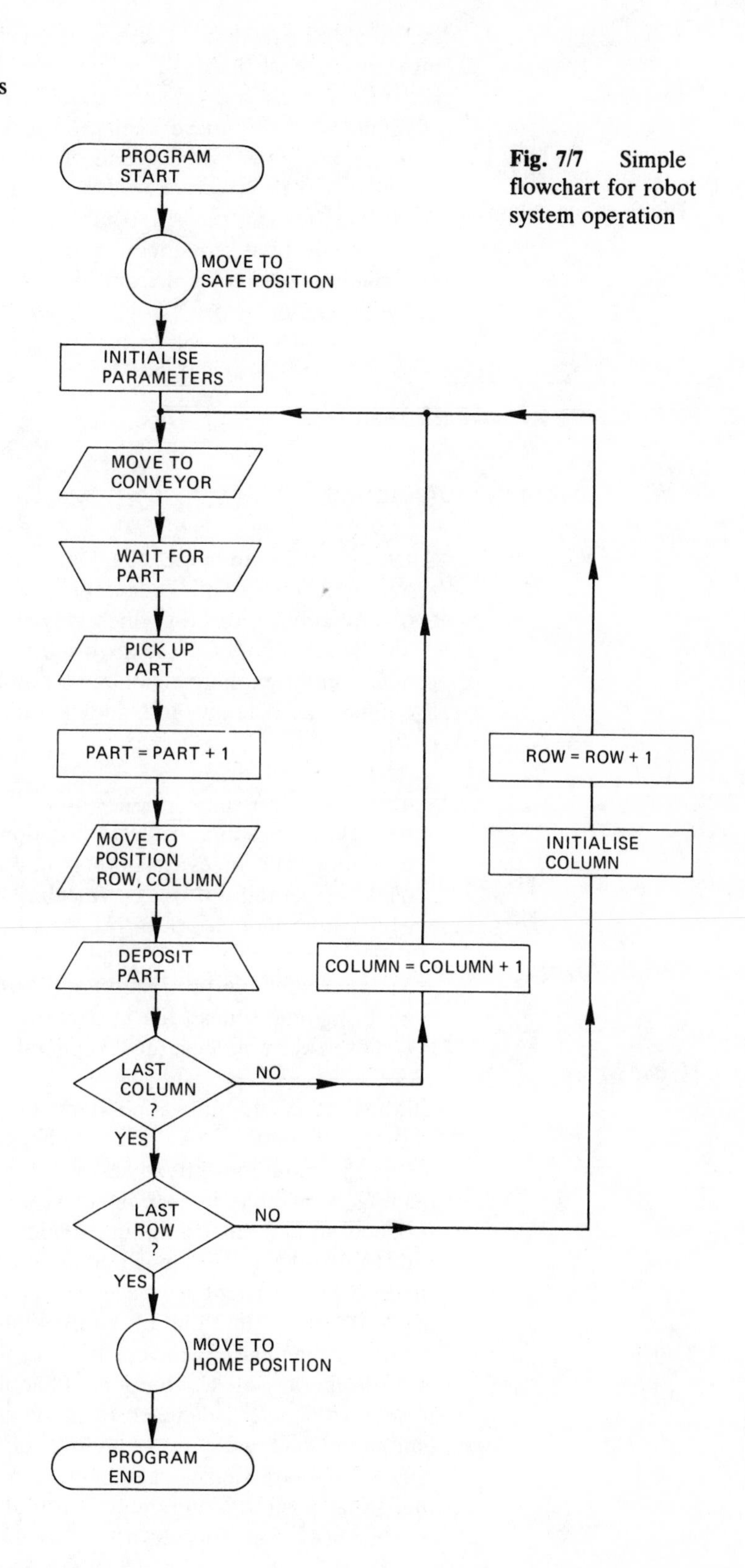

Fig. 7/7 Simple flowchart for robot system operation

identifiable operations. The operations should then be further broken down into their constituent sub-sequences. These sub-sequences can then be programmed as individual programs (modules) in their own right. The advantages are that these small programs are very much easier to comprehend than the whole task sequence in its entirety, and that they can be written and tested in isolation. Once every module has been tested, and is working correctly, they can then be combined to form the larger task sequence. Any errors in the final sequence can be tracked down to areas where operation effectively passes control from one module to another on the basis that the individual modules are error free.

7.3 Development issues

7.3/0 Flexible manufacturing

At this point in the book, you could be forgiven for regarding the robot as a system in its own right. The strength of industrial robots, however, lies in the tasks and applications for which they may be exploited, and not in the pursuit of the 'perfect' robot for its own sake. A modern approach to organising for production attempts to reconcile a number of factors previously responsible for much inefficiency and added costs during manufacture. Such factors include:

a) High (and rising) labour costs.
b) Labour discontent due to boredom, dissatisfaction, fatigue, injury, etc.
c) Labour shortages (in both skilled and 'hard-to-fill' areas).
d) Long setting-up times of machine tools (causing poor response time).
e) High levels of work in progress (due in part to *d*).
f) High inventory levels (due in part to *d* and *e*).
g) Unpredictable production and throughput times.
h) Long lead times (due in part to *d*, *e*, *f* and *g*).
i) Often variable accuracy, repeatability and quality of output.

The advent of computer-applied manufacturing technology now enables many of the above undesirable factors to be combatted and enables many organisations to become more flexible, responsive and competitive. The term **flexible manufacturing** has been coined to describe the resulting organisation for production. In metal machining applications, for example, Computer Numerically Controlled (CNC) machine tools are able to offer predictable throughput times, with consistent accuracy and repeatability, requiring short set-up times and with the minimum of human intervention. Key features of flexible manufacturing installations include the ability to sustain unmanned operation and automatic tool and component handling. Robots have a major role to play in supporting such advanced technological initiatives and as such have to be equipped to integrate smoothly into such automated environments. The key here is software that is capable of controlling and coordinating the various individual system components into a total integrated, automated manufacturing system. In support of this it is essential that robot control systems keep pace by offering the facility of communication with such system software. In addition to providing physical channels of communication (via input/output ports, etc.) it will become increasingly important to provide software access (programming facilities) to the robot operating system and sequence programs. In reality, a common approach is to automate and robotise self-contained applications within an already functioning manufacturing environment. Thus, 'islands of automation' are created in key task areas. The natural development is then, via software, to link the islands to form a coherent integrated automated manufacturing system.

7.3/1 Artificial intelligence

Artificial intelligence is the characteristic of a machine (almost by definition computer-based machines) to adapt to various conditions or changes in its environment. This should not be confused with the ability to sense conditions or changes in the environment (sometimes referred to as **machine intelligence**). For example, it may be comparatively simple to detect a change in temperature of the environment but it may be quite difficult to assess which one of a number of actions to take, especially if the particular temperature condition has not previously been encountered. The classic example usually portrayed is the chess-playing robot. Industrial examples are manifold, for example: parts or operation scheduling to minimise machine downtime, to minimise W.I.P. or to maximise productivity; component nesting to maximise the number of components accommodated and minimise free space; palletising or packaging components of different shapes or sizes to maximise the number of one type contained or to minimise free space and so on. Each task must be performed slightly differently depending on the desired outcome. One goal of artificial intelligence is to allow such tasks to be performed intelligently, in real time, depending on environmental conditions and under circumstances perhaps not previously encountered.

Components of artificial intelligence systems include:

1 *A means of representing knowledge and information*. This includes the representation of difficult concepts (consider the task of representing the theory of relativity for example), and representation to enable optimum storage and access. This resembles the human attributes of language and communication.

2 *A knowledge (data) base of information relating to known conditions, known solutions and the implications of known actions*. This includes researching improved ways of building, storing, searching and accessing large amounts of stored information. This resembles the human attributes of memory, recall and experience.

3 *Programmed strategies for utilising the information and knowledge available to make intelligent decisions*. Such strategies are termed **algorithms**. The decisions to be taken will usually be via programs to maximise or minimise some desired criteria. This resembles the human attributes of intelligence, reasoning, estimating and forecasting gained from, for example, experience, education and training.

4 *A means of expanding the knowledge base, based on newly encountered experiences and the decision-making process*. This resembles the human process of learning.

One means of providing a knowledge base upon which AI systems can be developed is to combine the knowledge, expertise and experiences of a number of respected practitioners in a particular field, with a statistical basis of analysis. Such systems are known as **expert systems** and are often used in diagnostic situations where certain symptoms can be identified and compared with the knowledge base. The probability of certain conditions can then be diagnosed and compared, providing the necessary information on which subsequent decisions can be based. Such software is already in widespread use, in engineering, in the form of CAD systems, engineering modelling software and diagnostic software.

In order to be able to extract the relevant information from experts, and represent it in a form that can be readily accessed, a new breed of computer scientist, the *knowledge engineer*, has been identified.

It can be seen that robots do not have the capacity for intelligence if they do not support sensors from which they can perceive and interpret their immediate environment. It should also be understood that it is not the sensors but the accompanying software and knowledge base that renders the robot 'intelligent'. There has already been much research carried out in many diverse fields seeking to emulate and improve upon the characteristics of human intelligence and performance. They include the following:

a) Image Processing and Pattern Recognition – (sight)
b) Sound/Speech Recognition – (hearing)
c) Voice/Speech Synthesis – (speech)
d) Tactile and Force Sensing – (touch)
e) Data/Information Storage and Representation – (memory)
f) Programming Languages incorporating Decision-making Capabilities – (consciousness and judgement)
g) Control Technology – (coordinated and controlled movement)

Artificial intelligence, still in its infancy, seeks to combine these with a **heuristic** (learning by discovery) approach. However, developments in AI software will need to be accompanied by parallel developments in programming techniques. Software is the vital link between the human and the robotic system and the facilitator of an integrated, automated and often unmanned manufacturing system.

7.3/2 The transputer

The developments of real time control of increasingly complex unmanned manufacturing systems, research into artificial intelligence and the likely requirements for ever more powerful robot vision systems are beginning to test conventional digital computing techniques based on the now familiar microprocessor. We have seen, in Section 6.3 for example, that even relatively fast microprocessors take inordinate amounts of time to digitise even medium-resolution images. Assigning a number of identical microprocessors to the same tasks will only be partially successful since the relative performance of these shared task systems degrades with the more processors added. Around 4 or 5 processors represent the optimum number. Conventional microprocessors operate, and are programmed, in a sequential manner which can in itself limit overall system response and operation due to local bottlenecks.

A new device known as a **transputer** is set to revolutionise computing techniques. These devices represent both an alternative hardware and software approach to computing techniques based on the perceived limitations and experiences gained using conventional microprocessors. Essentially a transputer (depending on the version) integrates a powerful 16-bit or 32-bit microprocessor, four or six high-speed data communication links, 2K or 4K of on-chip RAM, a memory interface for external RAM and peripheral interfacing, on a single chip. It is physically different from a conventional microprocessor since it is manufactured in a square package (rather than rectangular), and has 84 pin connections rather than the 40 pin connections of the conventional microprocessor. Tests confirm that it can outperform equivalent conventional microprocessors (in terms of processing speed) by between 3 and 10 times. Since it has the equivalent of some 250 000 transistors, the name is aptly coined from the term *trans*istor com*puter*.

The advantages claimed over conventional microprocessors include the following:

1 The hardware and software of the transputer have been designed together for optimum performance.
2 The instruction set of the processor has been reduced (to about 100 instructions) to increase speed of operation and ensure that the most-used instructions are executed most swiftly and efficiently. It is known as a **RISC** (Reduced Instruction Set Computer) device.
3 On-board RAM enables programs and data to reside within the chip and thus alleviates time overheads in accessing external memory (although this facility is provided if required).
4 Perhaps the most significant advantage is that it permits **parallel processing** to be accomplished. For example, in multi-transputer systems, different transputers can simultaneously execute the same code on different data, or different code on the same data. In single transputer systems, powerful multi-tasking processing is provided enabling different processes to be carried out almost concurrently.
5 The high-speed communication links (10 Megabits/sec) enable transputers to conveniently interact with each other and do not impose a limit on the number of transputers that may be connected together.
6 System performance increases linearly as transputers are added to the system. Thus the power of a system can be increased by simply adding more transputers.
7 The improved architectural design of the transputers enables their interconnection by only two wires per link, thus considerably simplifying overall system design.
8 A new high-level language (called *Occam*) has been devised to handle parallel processing. Since the transputer has been designed to optimise execution of Occam, it is claimed that the execution of compiled Occam is so efficient that low-level languages are unnecessary.
9 Occam is a highly structured language, enabling more efficient program development and the facility of integrating library routines on a modular basis.
10 Transputers can operate at higher cycle speeds than conventional microprocessors.

There is little doubt that such devices will become prominent features of digital computer based systems in the near future. They will require alternative methods of thinking concerning the application of their parallel programming capability and their integration into sophisticated systems.

Questions 7

1. Define 'lead-through programming' of industrial robots, discuss its relative advantages and limitations and state *two* industrial tasks for which it could be successfully employed.
2. Define 'drive-through programming' of industrial robots, discuss its relative advantages and limitations and state *two* industrial tasks for which it could be successfully employed.
3. Define 'off-line programming' of industrial robots, discuss its relative advantages and limitations and state *two* industrial tasks for which it could be successfully employed.
4. Briefly explain the concept of 'world modelling' in conjunction with the programming of industrial robots.
5. Define 'coordinate entry programming' of industrial robots, discuss its relative advantages and limitations and state *two* industrial tasks for which it could be successfully employed.
6. Describe the procedures that should be observed when commanding an industrial robot to return to its HOME position.
7. Industrial robot control systems have facility for setting various 'parameters'. Describe common parameters that:
   a) the user may alter, stating the reasons why
   b) are set by the manufacturers and the user should not alter.
8. Explain the difference between 'single step' and 'automatic' modes of robot operation.
9. Sketch a typical teach pendant, fully label each control function, and give a brief explanation of its function and purpose.
10. List the points to be observed during a typical startup procedure for an industrial robot.
11. Explain the relationship between 'ROM', 'RAM', 'sequence program' and 'control program' in the organisation of the computer control of industrial robots.
12. Differentiate between machine level, assembler and high-level languages as levels of programming language.
13. Describe *three* fundamental programming structures available in all robot and computer based programming languages.
14. Describe how the planning technique of 'flowcharting' can help in sequence program development by off-line techniques.
15. There are a number of specialised robot programming languages used in conjunction with industrial robots. Briefly outline the origins of such languages and their claimed advantages.
16. Outline the contribution of programming and software development in modern industrial trends towards flexible manufacturing systems.
17. Speculate on the role and development of 'artificial intelligence' techniques and their application to the future of industrial robotics.

Safety Considerations 8

8.1 The need for safety

8.1/0 Legal requirements

Both suppliers and users of industrial-type machinery and equipment (of whatever sort) are under a legal obligation to provide adequate safeguarding and safe working practices in respect of that equipment. Robots are industrial equipment and are thus subject to the same legal obligations and requirements. Failure to ensure safety can result in liability under both civil law, based on negligence and resulting in compensation, and criminal law, based on a breach of the law and resulting in punishment. In Great Britain the two major legislative Acts concerning, in general terms, safeguarding and safe working practices are:

1. The Factories Act (1961)
2. The Health and Safety at Work Act (1974)

The Factories Act covers general points such as cleanliness in the workplace, overcrowding, working conditions and other aspects of employee welfare. It covers aspects of safety in relation to work practice, guarding of machinery, materials handling equipment, fencing and other matters of access on the shopfloor.

The Health and Safety at Work Act is also general, rather than specific, and imposes obligations on both employer and employee. Essential provisions for the employer include:

- to ensure, so far as is reasonably practicable, the health, safety and welfare at work of all employees.
- the provision and maintenance of equipment and systems of work that are, so far as is reasonably practicable, safe and without risks to health.
- the provision of such information, instruction, training and supervision as is necessary to ensure, so far as is reasonably practicable, the health and safety at work of employees.
- to prepare and revise where appropriate, a written policy with respect to Health and Safety at Work.

Essential provisions for the employees include:

- to take reasonable care for the health and safety of themselves and others who may be affected by their acts or omissions at work.
- to cooperate with the employer so far as is necessary, to implement the Act.
- not to intentionally, or recklessly, interfere or misuse anything provided in the interest of health, safety or welfare in pursuance of the Act.

In devising physical and organisational controls for the safeguarding of industrial robots a common sense approach must be adopted. That such flexible equipment may not be able to be totally insulated from exhibiting potential danger must be faced. This possibility is established, in principle at least, within the above Act in the phrase 'as far as is reasonably practicable'. The absence of any specific or definitive guidelines often prompts responsible bodies to address such problems by independent research and analysis on behalf, and in the best interests, of their constituent members. Such investigation, usually between manufacturers, suppliers, users, government and academic bodies, may result in the production and subsequent publication of a code of practice.

8.1/1 Codes of Practice

A **code of practice** is a collection of advice, guidance, suggestions and recommendations on accepted methods of good practice and procedure on aspects to which the code relates. In general, codes of practice are not legally enforceable. They are drawn up by collaboration between respected professional, industrial, commercial, academic and government bodies based on technical knowledge, experience, observation, historical examination, statistical analysis, research findings, current good practice and plain common sense. The development of industrial robotics is still in its relatively early stages. This has a number of implications for the development of codes of practice relating specifically to industrial robot installations, for the following reasons:

a) Practical experience is at present limited, therefore definitive principles have yet to be formulated.
b) Basic principles of existing codes of practice, relating to industrial equipment in general, may not be entirely appropriate or transferable to robotic installations.
c) Industrial robots, and their application areas, are under such continual and rapid development that it is difficult to establish truly general guidance.
d) Future legislation specific to industrial robots may influence the guidance tendered in any codes of practice.

Two prominent codes of practice having relevance to industrial robots are:

1. British Standard 5304: 1975.
   Code of Practice: Safeguarding of Machinery.
2. Machine Tool Trades Association (MTTA): 1982.
   Code of Practice: Safeguarding Industrial Robots
   Part 1. Basic Principles Part 2. Welding and Allied Processes.

BS5304 Safeguarding of Machinery
This is a British Standard giving general guidance on methods of safeguarding industrial machinery. It indicates criteria to be observed in the design, construction and application of such safeguards, and how they should be used. The standard offers the guiding principle that, unless a danger point or area cannot be reached, then the machine should be provided with an appropriate safeguard which eliminates or reduces the danger, before access to the danger area can be achieved. Since it dates back to 1975, the standard does not specifically address the safeguarding of industrial robots. In this respect British Standards accepts that it is often impractical to apply the basic principles to the entire work area of, for example, robots. Complementary standards offering advice on specific applications are in the course of preparation.

MTTA Safeguarding Industrial Robots
The Machine Tool Trades Association is a central, national organisation charged with the task of furthering the interests of those companies who manufacture, or import, machine tools and allied equipment. The code of practice is published in two parts. Part 1, Basic Principles, describes the basic principles of safeguarding industrial robots other than mobile robots and master-slave manually controlled manipulators. It includes a systematic examination of the foreseeable safety problems which could arise as a result of introducing robots into the workplace. Part 2, Welding and Allied Processes, describes the additional hazards likely to be encountered when using robots in welding-related tasks including spot welding, gas-shielded arc welding, thermal spraying and cutting, hardening and welding by laser. This code of practice is just one of a large number of similar publications concerning safeguarding and safe working practice published by the MTTA.

In addition, the Health and Safety Executive (HSE) are very active and provide a number of useful working papers and guidance notes relating to specific aspects of safety across all disciplines.

8.1/2 Hazard analysis

The safeguarding of an industrial robot installation should be preceded by a thorough analysis of potential hazards and the source of those hazards within the workplace. This exercise will highlight those conditions under which the robot is likely to promote danger or inflict injury and suggest the extent of risk associated with each condition. When the likelihood and consequences of the potential hazard conditions have been assessed, the appropriate safeguards can be designed and applied. In many cases the assessment of risk and the identification of hazard conditions will be a matter of judgement and experience coupled with guidelines available within various legislative standards and codes of practice.

Hazard analysis should encompass those extraordinary conditions that may arise during periods when the robot installation is not working under normal production conditions. For example, surveys have highlighted that almost 90% of accidents concerning industrial robots occur during periods of programming and maintenance. Additionally those, often unpredictable, conditions that could arise due to loss of power or malfunction must also be very carefully considered and taken into account when designing safeguard systems. Another potential source of hazards may involve equipment necessary to the installation, but peripheral to the robot itself. Many robot actions may be determined or influenced by information feedback from external sensors or equipment. The mis-operation of sensory or external equipment must not be allowed to cause mis-operation of the robot itself to the extent that a hazard situation results. The hazard situations that may be encountered within a robot environment will include the following:

1 Hazards due to human failings, acts or omissions, including during periods of programming and maintenance.
2 Hazards due to designed robot operation and workplace layout.
3 Hazards due to hardware failure or mis-operation including loss of power supply.
4 Hazards due to control system failure or malfunction including software and data transmission errors.
5 Hazards due to the malfunction of external sensors, equipment and safety devices.

In all cases the provision of easily accessible emergency stop controls and the adoption of safe working practices and procedures should be prerequisite and regarded as of paramount importance. As part of a continuing philosophy of review, re-assessment and improvement, full documentation should be raised and maintained. Such documentation should include details of the analysis, decisions and systems of work finally implemented, together with an on-going log recording any maintenance, additions, modifications, malfunctions and other relevant incidents.

8.2 Safety hazards

8.2/0 Human considerations

The human groups at risk in a robot environment include operators, programmers, maintenance personnel, casual observers and others outside the assumed safety zone. Many potential hazard situations present themselves because of the failure of human personnel to appreciate the nature of robot behaviour within its working environment. The reasons are not difficult to identify and include the following:

1 Personnel tend to assume that the robot will operate and move correctly according to its programmed task. Malfunctions can cause serious departures from programmed operation, sometimes resulting in complete loss of control.
2 Personnel tend to assume that a working robot will continuously repeat an observed sequence of movements. Different programmed sequences may be activated dependent on the feedback from external sensors and equipment. There is thus a very real possibility and danger of inexperienced personnel being caught unawares if they are allowed uncontrolled access into the immediate robot workplace.

3 Personnel tend to assume that a stationary robot is a safe robot. There are many occasions when robots are programmed to remain stationary. Examples include waiting for external feedback or performing a timed delay or dwell. This can be signalled by means of (flashing) indicator warning lamps.
4 Since the extent of robot movement is usually well beyond its base dimensions, and is three dimensional, it is often difficult to visualise the safe zone around a particular robot configuration. The type and extent of robot movements and the shape of many working envelopes are often difficult for human personnel to visualise.
5 The speed of robot operation may well be greater than human reaction time, thus giving little time for evasive action to be taken.
6 Safe working practices and procedures may not be established, communicated or adhered to by all groups of personnel having cause to enter the robot workplace. Similarly, due to inexperience or unfamiliarity with equipment or otherwise, human operators are prone to making errors or errors of judgement.

A number of factors can be brought into play to reduce the risks outlined above. They include the following:

a) **Physical guarding and fencing**
Perhaps the most obvious outward sign that access to a robotised workplace is either restricted or prohibited is the presence of extensive guarding and fencing. The principle is that of demarcating and denying access to unsafe positions within the workplace. When such guarding is installed it highlights just how large the potential danger zone is. Guards or fences constructed especially to permit access (i.e. doors or gates) will commonly be interlocked to ensure their closure before correct operation can be initiated. An abundance of safety guarding must not promote a false sense of security. Examples of robot guarding and fencing are illustrated in Figs. 8/1 to 8/4.
b) **Intrusion monitoring to prevent unauthorised access**
In cases where it is impractical to erect physical barriers to prevent access, or as a double safeguard, intrusion monitoring is often employed. Intrusion monitoring means the detection of intruders. Various techniques are available. Photo-electric beams or light curtains, pressure-sensitive mats, infra-red and ultrasonic range sensors, and mechanical interlock systems operating on conventional guards or fences, can detect unauthorised access into the workplace. The systems are configured such that, if an intrusion monitoring device is tripped, the robot and associated equipment is halted in a safe manner. Resumption of system operation is done by restoring the intrusion monitoring system and consciously activating a re-start cycle. The choice of device will be determined by the extent of risk ascribed to the likelihood and type of access. Obviously, designs should be fail-safe and designed such that they cannot be bypassed either intentionally or otherwise.
c) **Safety control devices**
Devices such as pause/hold, dead mans control and emergency stop trip controls are examples of safety control devices. A pause/hold control stops operation of the robot whilst maintaining power to all axis and end effector functions. A dead mans control requires that it is held in a set position by an

Fig. 8/1 Guarding of a robot welding line [*courtesy: MTTA*]

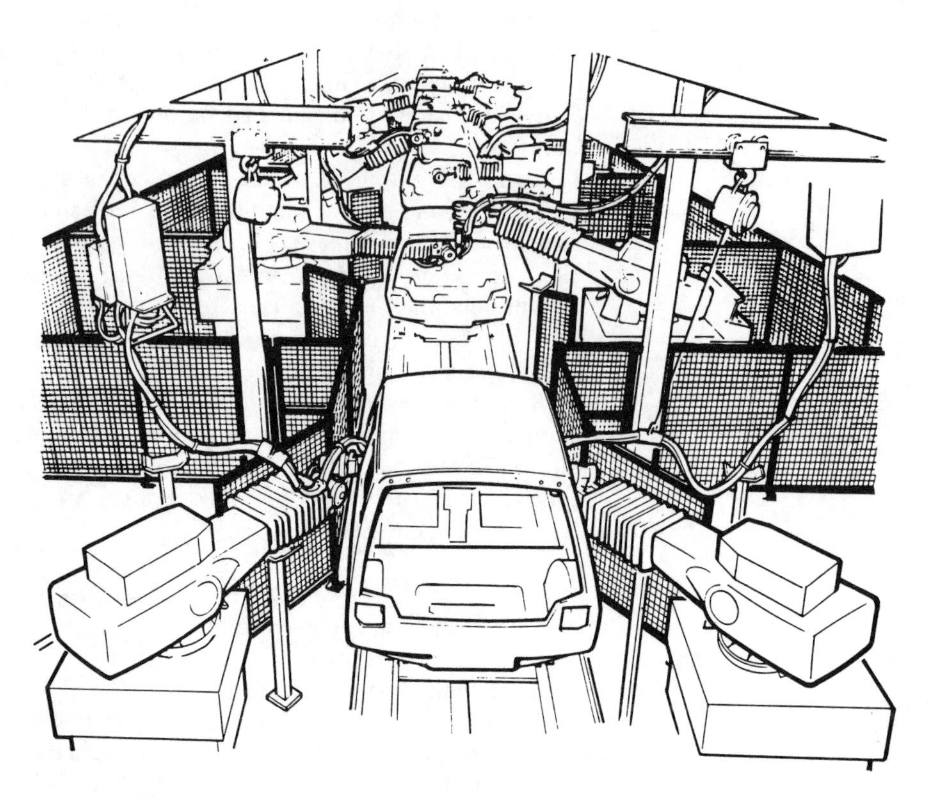

Fig. 8/2 Interlocked guard of robot machine tending [*courtesy: MTTA*]

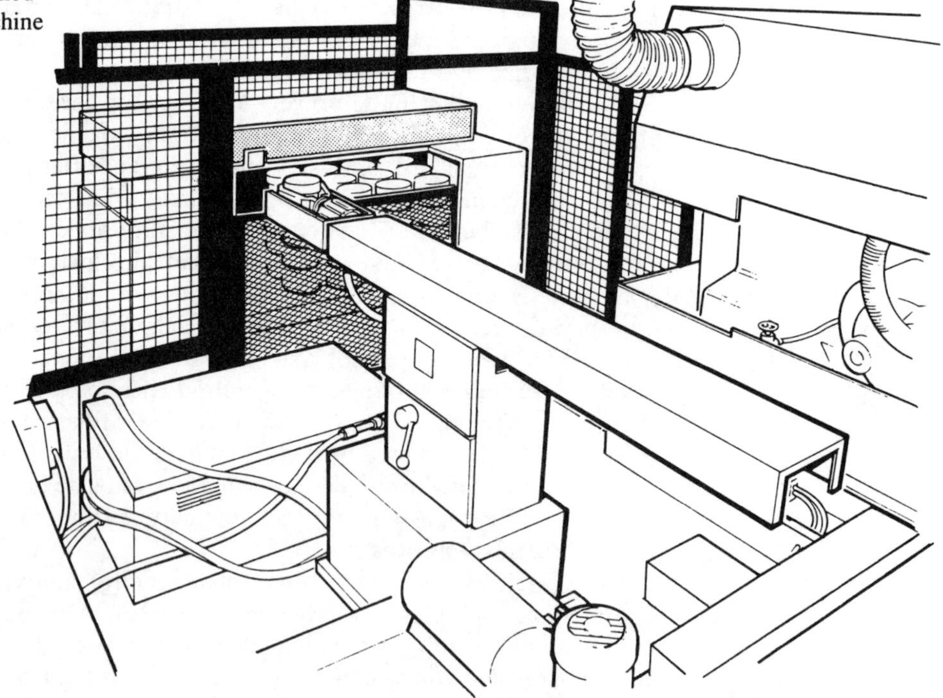

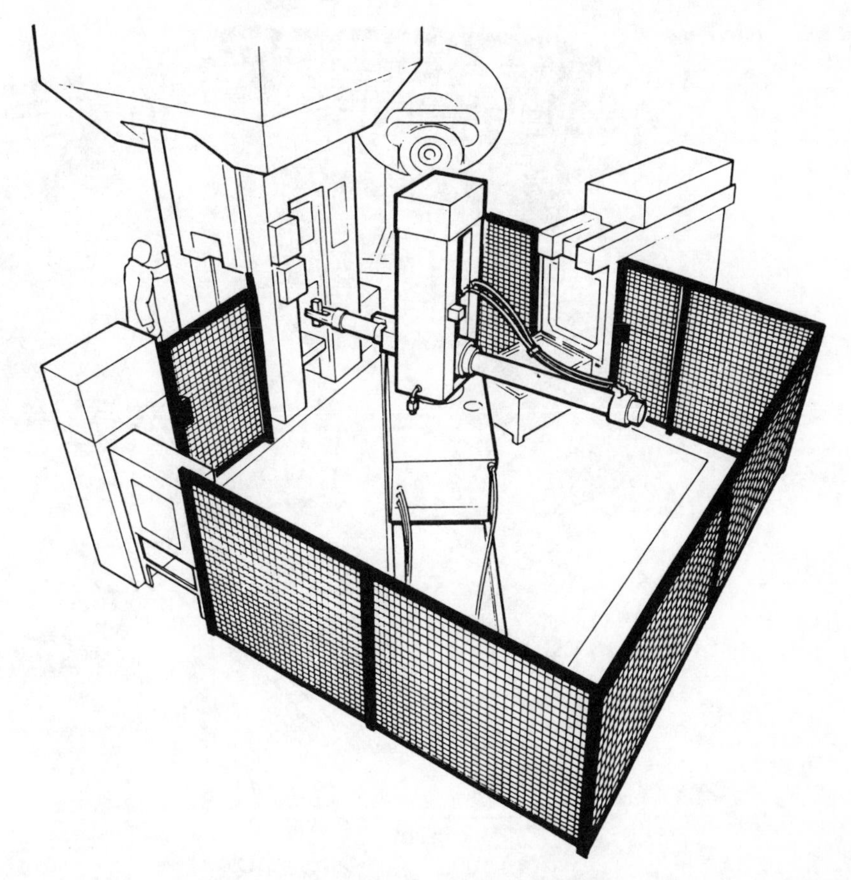

Fig. 8/3 Fixed and interlocked guards [*courtesy: MTTA*]

operator. If released, the control reverts to its original position and deactivates robot operation until it is again depressed. Emergency stop trips are similar to the type found in normal factory and workshop environments and on common machine tools and equipment. They should be prominently identified and within easy access of potential danger zones. Their action should be fail-safe and such as to disable operation in a safe manner.

d) **Safe systems of work**

Safe systems of work, including access into, and manual operation or programming of, robot environments, will minimise all types of hazard. They should be prominently displayed, thoroughly observed and meticulously documented. Oral instructions, requests or promises can be misheard, misinterpreted, misunderstood or forgotten and cannot be relied upon. A written permit-to-work system will exert control over those persons requiring access. Such permits must also be correctly cancelled by an agreed procedure. In the event of an accident, such documentation will form important evidence in discovering the cause of the accident and be instrumental in preventing a re-occurrence.

e) **Restrict mechanisms on robot operation**

In conjunction with point *d* above, teach-restrict and reduced speed and torque settings may have to be invoked to safely contain robot motion. This is especially important during programming, setting and maintenance operations.

Fig. 8/4 Fixed and interlocked guards [*courtesy: MTTA*]

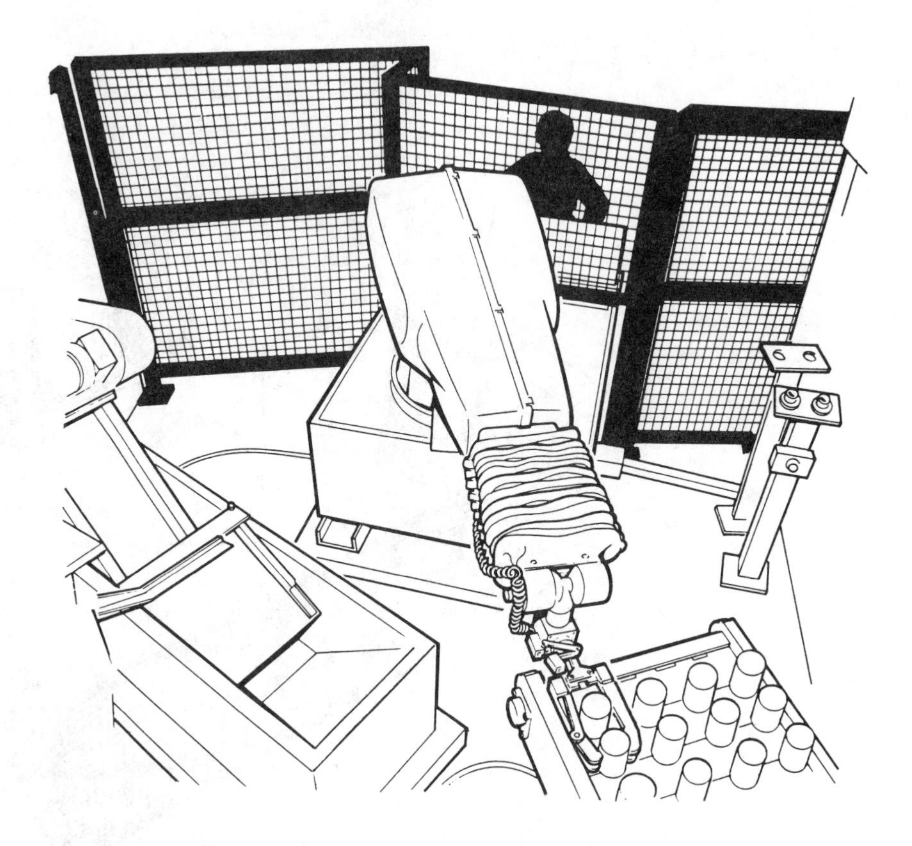

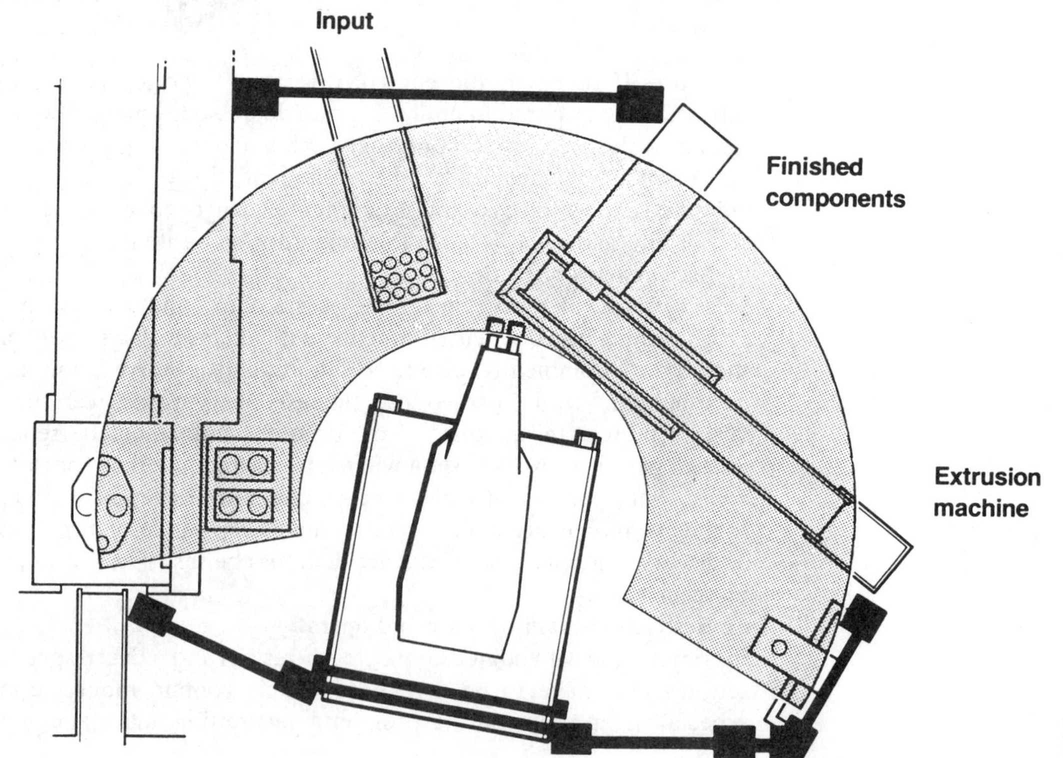

f) **Extensive training and familiarisation**
Perhaps the most important and cost effective safety prevention is that of extensive training, education and familiarisation of both robot operation and mis-operation, and the importance of observing strict safety procedures. All personnel either directly or indirectly involved with the operation of the robot and its associated equipment should be fully versed with the operation of both the robot and its control system. In particular the importance of safe systems of work and agreed safety procedures should be stressed.

8.2/1 Robot operation and the workplace

Even when a robot is operating correctly, hazards are continually present.

The first danger is that of *collision* between the moving robot and human operators. This potential hazard can arise through a number of circumstances. The most obvious are during programming and maintenance, but operators working in close proximity durings its normal operation are also vulnerable. Some programming techniques require that the programmer must be in close physical contact with the robot. Inexperience or human error may cause unintended and unexpected movement of the robot. Danger can be reduced by restricting the speed of operation of the robot axes. The danger when performing maintenance tasks is further enhanced by the fact that safety interlocks may have to be disabled in order to carry out functional checks.

A second danger is that of *trapping*. This can be reduced if fixed and rigid obstructions within the workplace can be minimised or eliminated. If fixed obstacles cannot be removed (e.g. in the case of a supporting pillar) then limit switches can be set to restrict the maximum allowable travel of axis movements. The use of dead mans switches may ensure that the operator is sufficiently removed from potential trapping zones. Some robots may be fitted with collision sensors on strategic points of the robot structure. Activation of these sensors will cause shut-down of the robot in a controlled and safe manner. Such sensors may also protect the robot from excessive damage.

Another potential hazard during normal operation concerns the dropping, throwing and mis-handling of *cargo*. Perhaps the most important contribution to this type of hazard is made by poor or inadequate workholding arrangements at the end effector. The cargo may be dropped or thrown due to a combination of bad grip, axis trajectory and excessive axis speed. The cause may be due to a stressed or poorly designed end effector, inexact programming, or presenting the cargo in such a way that effective pick up is at worst impossible and at best unpredictable. In the case of heavy or dangerous cargoes it may be necessary to totally enclose the robot working area by means of a strong cage.

8.2/2 Hardware failure and malfunction

Hardware malfunctions or power line/supply failures are among the most unpredictable yet most likely sources of safety hazard. The extent of danger will be dependent to some extent on the actuation system employed. *Electrical systems* present potential hazards from electric shock should a fault occur or the power umbilical become severed. Whilst power failure may immobilise the robot (designed inherently to fail-safe) it may also erase datum and program status information. The re-connection of power may thus cause unpredictable operation of the robot in an attempt to re-orientate itself or to resume a particular task where it left off.

Hydraulic and pneumatic systems are prone to immediate loss of pressure if fluid lines are severed or connecting unions damaged by collision. This can cause rapid movement of the robot structure and its cargo under the influence of gravity. Servo valves blocked or jammed due to a contaminated fluid supply can cause erratic and unpredictable behaviour. Mechanical failure can occur through overloading, corrosion or fatigue failure.

Whilst it is difficult to predict the failure of *physical components*, certain precautions can be implemented to lessen the likelihood of catastrophic failure: regular planned checking and maintenance procedures; the use of fail-safe devices; the use of duplicate systems; the use of physical safety elements such as steel wires to limit movement; and of course extensive emergency stop facilities.

8.2/3 Control system failure and malfunction

Any computer controlled system can suffer either electronic or software failures and malfunctions. Digital electronic control systems are inherently reliable since they are predominantly solid state (electronic with no moving parts) systems. However, excess heat, humidity, or voltage overload can cause damage or mis-operation.

Feedback signals from position and velocity feedback transducers, and environment sensors, can become corrupted either by failure of the transducers themselves or by electrical interference. This latter point is important in those applications where external currents and electrical radiations are likely to be significant, for example in arc welding applications. The other prime source of electrical interference concerns an unstable mains voltage supply. Supply spikes and troughs can result from momentary surges and overloads in the mains circuit where demands are being made by other equipment.

Software malfunctions are likely to arise from one or more of the following sources:

a) Data corruption due to transmission errors.
b) Inherent program or operating system faults (bugs).
c) Electrical loss or interference.
d) Electronic component malfunction causing inadvertent loss or corruption of memory contents.

It is for many of the above reasons that software interlocks are not recognised as being entirely safe. They also are difficult to check thoroughly since there may be a large number of possibilities, and problems may only occur under a combination of unique conditions. Safety inspectors cannot verify and certify them during on-site inspections.

A number of methods are available for combatting control system failures, in addition to routine checking and maintenance procedures. Many systems have resident diagnostic checking routines that can be invoked at any time; data transmission can be checked by a variety of means; battery back-up systems maintain the contents of volatile memory through a power loss situation; and system status and operation can be monitored by a computer system by continually applying signals to various points and receiving them back on a time-out basis. Data checking and verification during transmission is vitally important. It is possible, for example, for the process of data transmission to occur seemingly without error. However, without software checks during transmission, it would be equally possible for dimensional and/or command values to become altered during transmission.

8.2/4 External equipment

Although safety sensors, interlocks, cut-outs and other systems are widely employed, both on the robot itself and within the workplace, it should be appreciated that these components could also possibly fail. Regular, planned, systematic checking and maintenance procedures are required to verify the integrity of such systems. An alternative approach is to continually monitor the operation of important devices automatically under computer control. The concept of a 'challenger' which periodically activates the sensor(s) on a regular time basis is being widely employed in unmanned automated environments. The response of the sensor, to the challenger, is monitored by a computer system which can detect a valid or invalid response. A continuous cycle of 'challenger-sensor-monitor' operation is maintained during the normal operations of the robot.

Emergency back-up systems should be considered in all cases where solid state devices do not provide the assurance of reliability available with orthodox safety devices. Power failure or control system failure provide good examples. Furthermore, all such back-up systems should be regularly maintained, checked, tested and adequately protected.

Questions 8

1 State *two* major legislative Acts that have implications for the introduction of robotics into the industrial workplace, and outline their main provisions.
2 Outline the main responsibilities of employers in terms of health and safety at work.
3 Outline the main responsibilities of employees in terms of health and safety at work.
4 What are 'codes of practice' and how do such codes originate? State *two* prominent codes of practice that currently relate to industrial robotic installations.
5 What is the importance of the 'working envelope' when considering robot safety?
6 Discuss those factors of workplace layout and design that can contribute to operator safety, and suggest means of minimising such hazards.
7 Briefly discuss the modes of programming industrial robots from the point of view of safety.
8 Describe the process of hazard analysis and how it can be used to influence industrial robotic installations.

9 State *five* safety hazards that can be identified within industrial robotic installations.

10 Outline *six* factors, attributable to human failings, which can place human groups at risk within an industrial robot environment.

11 Briefly explain the principle of 'intrusion monitoring'.

12 Explain the concept of a 'safe system of work'. Outline your own thoughts and ideas in the design of a 'safe system of work' that could be implemented within an industrial robot environment.

13 Discuss the case for implementing 'preventive maintenance' to assist in raising safety standards.

14 Outline the reasoning behind providing physical protection, in the form of extensive fencing, guarding and caging in and around robot workplace installations.

15 Discuss the case for implementing a training and familiarisation programme on the grounds that it will contribute to greater safety. Outline the likely contents of such a programme, the likely duration and the likely client group.

16 State *two* aspects of robot operation that are most likely to provide immediate hazards to operator safety. Suggest how the hazards can be minimised.

17 Identify likely hardware failures and malfunctions within a robot workplace and suggest means by which such hazards may be reduced or eliminated.

18 Identify likely sources of robot control system malfunction and suggest how they may be reduced or eliminated.

19 Given that safety devices and sensors can also fail or malfunction, suggest ways in which such malfunctions can be quickly identified or minimised.

20 As safety officer in an organisation about to introduce industrial robots, produce a plan of action (complete with suggested timescale) concerning all aspects of safety and its implementation. Suggest how you would monitor the successful operation of the plan.

Economic and Social Implications 9

9.1 Economic considerations

9.1/0 Elements of cost

The decision to introduce a robot into the workplace must be carefully analysed since the potential costs of introduction go far beyond the initial capital purchase. We shall see too that the potential benefits to be gained may be more far-reaching than the figures first suggested in the simple calculation of a payback period. The elements of cost that must be considered in any robot installation include the following:

***a*) The capital cost of the robot**

Perhaps the most obvious element of cost is that of the robot manipulator and its associated computer control system. It should be clear, however, that the presence of the robot alone is unlikely to be capable of performing an industrial task without considerable hardware, software and human support. It is also interesting to note that although the costs of computer systems continue to fall, and the technology of the physical component parts of robots are well established and proven, the costs of robot systems themselves are tending to rise. Factors such as physical size, payload capacity, manoeuvrability, control capability, accuracy and repeatability all influence the initial capital cost of the robot. For these reasons it is important that the immediate tasks to be performed by the robot, and the future demands likely to be made on it, are closely analysed and the correct choice made. This will prevent a robot system being over-specified and hence being unnecessarily costly, and also the possibility of the robot system becoming inadequate or redundant as future requirements or developments change.

***b*) The cost of the end effector**

It is the end effector that transforms a programmable manipulator into an industrial robot capable of performing a useful task. Unless a turnkey robotic installation is specified, *standard* robot/control system combinations are invariably priced without consideration of the end effector. The purchaser has no option but to provide a *suitable* end effector unit in order to render the robot useful, but if a *specialist* end effector is required, the final costings may also have to embrace design, development, manufacturing and testing costs. The costs of the end effector may thus represent a significant proportion of the total cost of the installation.

c) **The cost of additional support equipment**

In many cases the robot is employed in tasks involving parts that have to be presented to it in a particular position, with a particular orientation, at a particular time, sometimes with some kind of relative movement. In such cases additional equipment may be necessary to transport, position, orientate or move the components in question. This comes under the general heading of parts presentation. Considerable time and cost may have to be invested in sorting and feeding devices or in providing conveyor systems, fixtures or mechanical manipulators. Offset against this may be the fact that much of the support equipment may indeed have been necessary to support the task had it been carried out by human labour.

d) **The cost of sensory feedback devices**

If there is variability present within the anticipated tasks then sensory feedback devices may be required in order to enable the robot to perform the tasks successfully. The sensors themselves may need additional hardware support in terms of interfaces, terminals, etc., and additional software to enable their compatibility with the robot control system.

e) **The costs of safety provision**

In its simplest form, safety provision includes the fencing off of the immediate robot environment. This may be further supported by intrusion monitoring systems, interlock systems and/or the development of safe systems of work requiring administrative and procedural back-up.

f) **The cost of installation**

A significant part of total robotisation costs are the costs of installation and commissioning of the robot workplace. The costs of installation can accumulate from many sources. For example, the robot itself may have to be mounted in a different arrangement. This may range from a simple pedestal or base to a more involved overhead-type gantry. More complex arrangements can include movement of the robot and some means of locomotion. The robot will need access to basic services such as electrical, hydraulic or pneumatic supplies. Elements of the workplace, such as conveyor systems or machine tools, may need siting or re-siting to be within the working envelope of the robot. Sensor devices may need installing remote from the robot and require routing to the control system. Sensors (notably vision systems) may require special working conditions to be arranged in order for them to function successfully.

g) **Cost of maintenance support**

Robots and their associated end effectors are high technology equipment that draw together a number of different disciplines. They are also employed in routine but often key positions within the total manufacturing system. For these reasons they are likely to require a responsive, high-calibre and multi-disciplined maintenance support function. These costs must also include regular overhaul and servicing costs. There will be a requirement for the digital control element of the robot system to be accommodated and this may extend to low-level software expertise in addition to microelectronic expertise. Thought must also be given to the robot becoming temporarily disabled for an extended period of time.

h) **Cost of programming**
Robots, like computer systems, are impotent unless they have sequence programs to execute. Industrial robots are relatively straightforward to program and this may be accomplished by a number of methods. Depending on the number and range of tasks for which the robot is employed, and the method of programming, it is likely that programming, attendant editing, and program documentation will be carried out by a specialist programming function within the organisation. Thought must also be given to the fact that some modes of programming require the robot to be taken out of service for the duration of program development. If the robot is permanently sited, it may also mean that part of the manufacturing line to which it relates may be similarly disabled.
i) **The cost of 'knock-on' effects**
Robots are immensely successful in highly organised and predictable tasks. However when robots do replace human operatives in the performance of a particular task it becomes (sometimes painfully) clear just how adaptable human operators are in exercising manual skills. Human operators can continually compensate for unexpected errors and variability in any area of the task being performed. Further, they can exert acute judgement in assessing the quality of the components used and the success of the task performed. This sense of adaptability and judgement is lost when a robot is applied to the same task. It may therefore be necessary to tighten up on the quality of certain aspects of incoming components to the robotised task, which previously could be compensated for by the human operator. Additionally, re-design of the components to facilitate production via automated methods may also be desirable. Although this will undoubtedly contribute to a better product it may take considerable time, effort and cost to identify, and initiate.
j) **The costs of insurance**
Most manufacturers and suppliers of advanced and high technology equipment usually levy quite high insurance charges on the hardware and particularly the software which they supply. The nature of the product makes it likely that an in-house maintenance facility cannot support such equipment, and certainly not the software. Since the nature of such an investment means that the company purchasing the equipment will place a high reliance on its operation, it is unlikely that it will refuse such insurance cover. Insurance rates are usually quoted on a percentage basis (10% being a typical figure) which is a significant additional running cost.
k) **Cost of staff training and awareness**
Costs of training staff vary with the number of staff directly involved in supporting the robot installation. A managed system of training will need to be considered. This may embrace management personnel, direct supervision, programming, maintenance and production engineering personnel, and members of the workforce directly affected by the installation. Quite apart from the many technical aspects of the installation, a much wider awareness programme concerning (at least) safe working practices and procedures are also likely to be necessary.
l) **Cost of investment**
The costs of the total installation must be carefully weighed against the costs of carrying out the tasks via other methods. Using a human operator is the obvious parallel. It should also be realised that the cost of raising the capital

for the installation must also be considered. For example, if a loan has to be raised against the capital costs incurrred then there will be interest charges. Regardless of the source of the capital, the investment potential of the capital sum must also be considered. This is the amount of interest the capital investment would accrue if it were invested to realise a more secure rate of return. Superimposed on these considerations is the fact that the robot is likely to depreciate in value as its service life increases.

In addition there will be the running costs of the installation to consider although these are likely to become insignificant in comparison with the other elements of cost, and the running costs of supporting human operatives.

9.1/1 Justification of robot installations

The impetus for considering the introduction of robots into the workplace may arise from a number of commercial influences. These include:

1 Inception of new products, processes or factories.
2 Part of a management-led business strategy.
3 Production engineering or economic problems (e.g. quality, productivity or manning difficulties)
4 Customer pressure (notably from large organisations or government agencies).
5 Competitive influences within the marketplace.
6 Vendor (robot manufacturer) pressure.
7 Prestige or the nature of the business (e.g. aerospace, defence, nuclear and other high-tech product areas)

It can be seen that the origination of robotisation proposals can be from the top of the organisation down, or from the lower level upwards. The source is likely to influence the way in which it is received throughout the organisation. It must be realised that it may need much hard work and conviction to enable robotisation proposals to come to fruition. There must be commitment from top-level management to ensure that the impetus is maintained; there must be commitment from middle management for its effective management; there must be commitment from technical and production engineering staff to ensure that the system works effectively; and there must be conviction from the workforce to readily accept its introduction.

Unless the initiative for robotisation originates from top management it is likely that all proposals will require substantial justification on financial and economic grounds. The traditional approach to such justification is the concept of a *payback period*. A payback period is a duration of time during which the combined benefits derived from installing and operating the robot are deemed to have covered the costs of doing so. The robot is then said to have paid for itself and thereafter is operating in a profit situation. This means that the benefits used to calculate the payback period now contribute largely to net profit. To be considered on these grounds a typical robot installation would require a payback period of around three years. A simple formula for calculating such a payback period is

$$P = \frac{T}{S - R}$$

where P is the payback period in years
T is the total costs of the robot installation
S is the annual savings made by operating the robot installation
R is the annual costs of operating the robot installation.

This somewhat basic formula can be modified by other relevant quantities identified as exerting a prominent influence; for example, the effects of wage rate inflation, of productivity measures, quality and re-work measures etc.

History and experience may eventually suggest that such traditional approaches are not entirely appropriate to many robot and other advanced manufacturing technology (AMT) installations, for a number of reasons:

a) Payback periods concentrate on short-term benefits and many robotic investments probably offer more lasting, flexible and beneficial advantages as a long-term commitment.
b) Complex robotic installations are likely to have an installation period of a number of years which may well exceed the payback period demanded by traditional payback accounting methods.
c) Many benefits may be apparent but not easily quantifiable in strictly financial terms. For example, reputation, morale, knock-on effects in other facets of the business, savings in floor space and so on.
d) The costs of robotic and ancillary equipment is extremely high and cannot often be discounted against purely direct costs such as labour savings, reduced inventory levels, higher productivity, etc. (which ignore the wider benefits that may be derived), within a reasonable payback period.
e) Purely financial measures do not account for the flexibility of robotic equipment to respond to changing circumstances. For example, robots can be switched on and off as desired, can be moved from task to task, do not present hiring or recruitment problems or costs, and alleviate damaging industrial relations problems at often crucial times.
f) The benefits to be gained may increase as the level of advanced manufacturing technology increases to a situation where different elements can integrate under computer control and coordination.

Research has suggested that, on taking account of all cost factors, a 'robot wage' should be established and used for comparison with operating a human workforce.

Traditional accounting methods are likely to continually underestimate the potential advantages and benefits of robots, especially under payback period calculations. It is important therefore that some creativity is exercised in presenting the case for a robotisation project. Perhaps the most convincing argument is a demonstration of the proposed robotic system in action. All interested parties can then witness, first hand, the potential benefits and advantages (together with any attendant disadvantages), offered in making such an investment. Case studies of existing installations and evaluations offer a second avenue to be explored.

9.2 Social considerations

9.2/0 Integrating robots into the workplace

In the majority of robot installations human operators work alongside, and sometimes in conjunction with, robot devices. Mention has already been made of the need for consultation at all stages of a robotisation program, and also the need for substantial education and training to support its implementation. Certain social and economic factors, many perhaps not yet fully encountered or appreciated , may prove to become significant in the circumstances of robot/human interaction. For example, robots are tireless machines and as such do not communicate or interact socially. This may tend to bring about a certain alienation of the human worker. The tireless and 'perfect' nature of the robot's work may become a source of resentment to those with normal human failings. The robot can quickly learn skills of human dexterity, learnt over many years by time-served operators. Furthermore, robot skills can then be saved for later use. These factors can contribute to a further resentment, a sense of being used, a feeling of worthlessness and ultimately a demoralising loss of self-esteem. Since robots are capable of being closely measured and monitored, human operators working in conjunction with robots may feel that it is they themselves that are being monitored, measured and compared.

Since robots are employed in physically demanding jobs, there may be a tendency to increase the capability of the task to a level that can easily be accommodated by the robot. In such cases an increase in productivity will almost certainly be measurable. However, should the robot fail it may also be necessary to re-deploy human labour to carry out the task. The implications are that, first, human operators may not be available since robots may have displaced them and, secondly, that the task may now be beyond the physical capabilities of the human operator.

The integration of robots into the workplace must be kept in perspective. Because of the rapid, accurate, repeatable, tireless and reliable nature of robot devices, it is often a failing of (often influential) human observers to attribute human qualities to them. At worst this failing may lead to the notion that human operators are expendable. At best, human operators may not become recognised for their unique qualities of wisdom, experience, judgement, worthiness and those other assets that make humans such a valuable resource.

9.2/1 Robots and management

Management of manufacturing organisations have certain fundamental economic and commercial responsibilities. For example, they have to:

a) Ensure that the organisation remains in business.
b) As far as possible, take steps to ensure that the organisation returns a trading profit.
c) Make efficient use of all the resources at their disposal.
d) Minimise operating costs.
e) Invest wisely for the future benefit of the organisation.
f) Strive to maintain a competitive position in the marketplace.

g) Ensure that the customers' needs are met in terms of specification, quality, quantity and delivery dates.
h) Be responsive to changing products, markets, needs and demands.
i) Be receptive to both the short-term and long-term needs of the organisation in particular, and the industry in general.
j) Maintain a smooth and stable operation.

Manual versus robotic systems will be compared and evaluated by measures which reflect those responsibilities. It is important to management, therefore, that manufactured products are produced at a predictable rate, with predictable quality and at a predictable cost per unit consistent with 'reasonable' investment and running costs. To management, robots conjure up (rightly or wrongly) strong feelings of:

a) Organisation and efficiency.
b) Tireless and uncomplaining labour, always available.
c) Predictable, reliable, consistent and efficient output.
d) Controllable and costable performance.
e) Flexible and adaptable utilisation, at short notice if required.
f) Long useful life with low overheads.
g) Unceasing performance even under adverse working conditions, and in hard-to-fill occupations.
h) Undisputable loyalty and industrious application.
i) Absolute control over production rates, WIP, stock levels, etc.
j) Economic use of floor space, consumables, etc.
k) Prestige, satisfaction and high morale.

It is probably true that, from a management perspective, the positive effects of robotisation tend to be over-emphasised and the negative effects somewhat under-stated (no-one wants to admit making a wrong decision to invest!). Conversely, experience has shown that human workers, often only seeking fair treatment and reasonable conditions, tend to suffer the reverse criticisms (not without some justification in some instances!). It is not difficult to appreciate that robots do present an attractive economic and commercial option for management to consider.

9.2/2 Robots and the workforce

The needs and expectations of workers are different from (but not necessarily incompatible with) those of management. In return for a loyal and industrious contribution to an organisation, most workers seek certain rewards and benefits. These rewards and benefits may include some or all of the following:

a) A fair remuneration.
b) Safe and congenial working conditions.
c) Security of tenure (job security).
d) Fair treatment and consideration.
e) Rewarding and challenging work.
f) Responsibility and an opportunity to contribute.
g) Prospects of promotion and advancement.
h) Self-esteem and recognition of their value (job satisfaction).
i) A share in the success of the organisation.

To many manual workers the prospect of robotisation may initially pose the threat of job losses. They may also be mindful that the robots may be applied to those tasks that robots can accomplish best, leaving other jobs to be allocated to the human operators. Whilst much publicity advocates that this will usually enhance the working standards of human operators, it may in certain instances result in the reverse situation. If such fears can be allayed then robots are usually accepted as a welcome and useful addition to the range of equipment available. There are many reasons why manual workers should approve the introduction of robots into the workplace. Under certain circumstances, robots can:

a) Perform boring, repetitive and mindless tasks.
b) Carry out heavy and arduous work.
c) Work in dangerous, unsafe or obnoxious environments.
d) Reduce the pressures and stresses of working to strictly paced production targets.
e) Eliminate those tasks that tend to isolate human workers.
f) Contribute to an increase in productivity from which all can benefit.
g) Release human operators, previously employed in undesirable tasks, to be deployed in more rewarding and satisfying work.

There are many ways in which robots can be accommodated without recourse to large-scale redundancy programmes. Planned labour rationalisation may be brought about in a number of ways. For example:

a) Displaced workers can be redeployed in other areas, possibly in the many support areas required by the robot installation itself.
b) Natural wastage of personnel leaving the organisation.
c) Re-training instead of recruiting for new job opportunities.
d) Retirements without replacement.
e) Voluntary redundancies.

From the perspective of the human worker the subject of robotisation is an emotive issue. It is important therefore that all relevant factors are discussed openly and considered objectively, and that available options are explored. No-one is likely to benefit from entrenched attitudes or autocratic dictatorship.

9.2/3 Robots and the economy

To manufacturing industry, robots offer a promise of reliability, controllability, measureability and predictability. These are essential ingredients in planning and operating successful manufacturing operations. The above attributes however, are not achieved cheaply since the creation of complex software and communication links between the robot(s) and other elements in the manufacturing chain is expensive to develop, implement and maintain. Manufacturing installations that are purpose-designed and created with integration in mind have a greater chance of sustained, successful operation within a reasonable time scale than those attempting the integration of robots with existing equipment. Since the latter is the situation most likely to be encountered for the foreseeable future this indicates quite serious economic implications for both manufacturing industry and the robot system manufacturers and suppliers. This may account for an eventual downturn in the growth of the robot economy itself and a suppression, by default, of the manufacturing industry which it would naturally serve, relative to competitor economies.

A second economic perspective which mitigates against robotisation is the fear that it necessarily displaces jobs. There is still little hard evidence to suggest however, that the natural progression of robotisation is the totally automated factory and the consequent armies of unemployed factory personnel. In reality it is more likely that the emergence of robot technologies is creating more jobs than it is causing to be shed. Consider the many robot manufacturing companies that have been created and the associated back-up (sales, research, installation support, servicing maintenance etc.). Jobs being lost in some sectors of the economy (predominantly manual tasks in manufacturing industries) are being created in those sectors previously mentioned. There is a shift in the deployment of jobs.

A consequent shift in the skills required of workers entering or transferring within manufacturing industries has been identified. Subsequent skills shortages within manufacturing industries will have implications for the education, training and re-training mechanisms that operate within the economy. It is likely that craft skills will gradually become replaced by technician-type skills in production engineering, systems design and building, multi-disciplinary maintenance skills, fault tracing and diagnostic skills, keyboard and programming skills and so on. This 'skills shift' phenomenon is not new and occurs fairly 'regularly', although on a somewhat large and protracted time scale. (For example, consider the emergence of the motor car and the demise of horse-drawn modes of transport, and the effects of the industrial revolution.) Socially and politically, however, it creates periods of instability that can often be difficult to deal with and places responsibilities on both government and industry to respond positively. It is essential that the economy is geared to providing an availability of competent technicians able to support emerging advanced manufacturing technologies.

In economic, social and political terms, the industrial applications of robots must be vigorously encouraged since the following benefits will undoubtedly be derived:

a) Cheaper commodities
b) An increase in quality
c) More efficient production and supply
d) Reduced materials and energy costs
e) The emancipation of human labour from unsafe, mindless and arduous work in often poor working environments
f) An improved competitive position in the national and international marketplace.

Socially it would appear that the benefits, easily identified and obviously derived, from employing industrial robots are not yet adequately or fairly apportioned. For example, if it is accepted that the nature of automation in manufacturing industry is likely to displace jobs and increase unemployment, then some of the benefits derived from such large-scale automation could be directed to easing the situations it helps to exacerbate: by, for example, initiating re-training programmes for displaced workers or providing the means whereby they may gain acceptable employment in other areas. Politically, the problems appear to be essentially two-fold: identifying those courses of action likely to provide an acceptable solution to the problems, and deciding on where the responsibility should lie for their implementation and funding. Although one common argument for robotisation is that it will free large sectors of the working population from the drudgery of mindless work and enable them to pursue opportunities for enriched leisure, this seems to oppose the traditional attitude to work. A fundamental change in this work ethic may encourage the acceptance of leisure activities as being gainful. The alternative suggests a possibility that leisure time will be just as unfulfilling and mindless as the mindless work it has replaced.

Questions 9

1. Categorise the costs of robot operation in terms of 'fixed costs' (those costs that are unaffected by the productivity of the robot), and 'variable costs' (those costs that vary with the operation of the robot).
2. Define the term 'turnkey project' in the context of a robot installation.
3. Explain why the cost of the end effector is normally a substantial proportion of the cost of a robot installation, over and above that of the capital cost of the robot manipulator itself.
4. In addition to the cost of the robot manipulator and the end effector, identify *four* other elements of cost that relate solely to the immediate workplace environment.
5. Identify and discuss possible extra knock-on costs that could materialise as a direct result of robotisation.
6. Suggest *six* reasons by which the introduction of robots into the workplace may be justified.
7. Discuss the concept of a 'payback period' in the justification of introducing robots. Define how such a payback period may be calculated.
8. Suggest *five* reasons why the concept of justifying robots by payback period alone may not be fully appropriate.
9. Identify and discuss areas of short-term economic benefit that could be derived from the installation of industrial robots.
10. Identify and discuss areas of long-term economic benefit that could be derived from the installation of industrial robots.
11. Outline possible problems that could be encountered in attempting to integrate robots into a manned industrial environment.
12. Present a case, on behalf of management, justifying why robots should be introduced into the workplace.
13. List *six* fundamental expectations of shop floor employees.
14. Discuss the possible advantages to shop floor workers of introducing robots into the workplace.
15. Present a case, on behalf of shop floor workers, justifying why robots should NOT be introduced into the workplace.
16. Discuss *five* ways in which planned labour rationalisation could be accomplished in order to facilitate a robotisation programme.
17. Discuss the long-term effects of robotisation (whether implemented or not), on company, local, national, and international economies.

Assignments 10

10.1 Robot applications

10.1/0 Application case studies

It can be argued that the majority of manufacturing engineers become involved in industrial robotics because of what they can do rather than what they are. For example, in 99% of robot applications, it is likely to be more important that a robotised operation can increase productivity, improve quality and reliability, or bring about economic savings rather than the fact that it has a jointed arm configuration, is operated by electric servomotor drives, and can be programmed by a number of techniques. Although the latter may be important in their own right, the selection of an appropriate robot system must start with the application concerned. The following case study assignments illustrate conditions under which the application of industrial robots in a manufacturing environment would be considered. Each case study depicts an industrial situation and requires the application of original thought in completing the assignments. Information research and retrieval should form an integral part of each assignment. The contacting of manufacturers and industrial suppliers for information, and the reading of trade journals, should be encouraged as should the competent presentation of both written and oral project reports. In all cases where information is omitted or incomplete, sensible assumptions may be made for convenience. These assumptions should be clearly indicated and should be justified by supporting information where appropriate.

10.1/1 Case Study 1

Application A firm produces various moulded plastic dashboards for use in the motor industry. The products range from simple laminar sheet mouldings in plastic to thicker 'padded' moulded foam plastics. The dashboard panels are moulded as a complete dashboard housing and the apertures cut out in a subsequent operation.

Present System The apertures are cut out manually using hand tools which, although representing only a small investment, present the following deficiencies:

a) The process is inherently slow since the materials are relatively difficult to cut; they are variable depending on the vehicle model; the aperture patterns frequently change due to changes in model and trim specifications; and cutting aperture details must be accomplished in three dimensions.
b) Production rates and throughput times are variable and difficult to predict with any accuracy.
c) Two operators are engaged on the task principally to reduce handling difficulties.
d) The process produces harmful dust, presenting a dirty and hazardous working environment to the human operators.
e) Accuracy of the manually cut apertures is variable, causing an unacceptably high rejection and re-work rate.
f) The finish of the edges of the cut apertures is variable which often necessitates a further finishing operation.
g) Established aperture patterns are likely to change as models or trim levels change.

Alternatives As a means of overcoming the above deficiencies a number of alternatives are being considered.

1 *Moulding in the apertures during the moulding process*
This would eliminate the need for a second operation, but would probably be the least flexible alternative. Since various apertures have to be provided to accommodate different models and varying levels of instrumentation, this alternative is likely to:

a) increase the complexity of the tooling
b) require large investment in different tooling arrangements
c) require long changeover times during product changeover or large stocks of moulded products
d) present handling/unloading difficulties due to the reduced rigidity of the perforated moulding.

2 *Producing the apertures by a punching operation*
This alternative would stabilise throughput times, render the accuracy of the apertures consistent, and eliminate the environmental hazard of harmful dust. It presents the following drawbacks:

a) Manufacturing on-costs per component will be high unless long production runs of identical components can be guaranteed.
b) Complicated details may not be technically feasible by this method.
c) Punched edges will produce burrs which require finishing.
d) New aperture details will require new tooling or extensive modifications to existing tooling.
e) If the product line ceases the investment is likely to be rendered redundant.

3 *Employing a robotised water jet cutting station*
This alternative will produce accurate, high-quality apertures of complex three-dimensional form at predictably high output rates. Aperture changes can be easily accommodated and the system could respond quickly to all component and material variants fed to it. The system could adapt readily to changes in production procedures and would immediately eliminate the direct costs of two human operatives. The system would require a large initial capital investment but this would be offset by savings in direct labour, increased productivity, reduced reject rates, reduced quality costs, and the potential offered for linking to other automated manufacturing systems. Other less-tangible benefits may include increased customer satisfaction and confidence and the fostering of an image and reputation for consistent, high-quality and reliable work. The robot manipulator could easily be re-deployed in quite unrelated tasks should the product line cease.

Assignment Tasks Assuming that the decision is taken to follow option three:

1 Suggest any problems (technical or organisational) that could be encountered in the installation of the robotised water jet cutting system and identify the support services and equipment that will possibly be required to render the robotised water jet cutting station operational.
2 Draw up a detailed specification for a suitable industrial robot that would be capable of performing this task. You must supply support information concerning the likely sizes of the moulded components and the reach and manipulative characteristics required. Compare your specification with the example specifications that appear at the end of this chapter and state which of these, if any, would be suitable. State, giving supporting reasons, the mode of programming that you would suggest for this particular application.
3 Compile a detailed plan, including time duration projections, of how you would approach the installation of this project. Your plan must accommodate physical, organisational and human factors. State your projection of how long the installation will take to complete.

10.1/2 Case Study 2

Application A new manufacturing unit is to be set up to automatically produce a family of similar axle shafts. It has been decided that a self-contained flexible machining cell comprising a number of machine tools, tended by one or more industrial robots, must be designed to accommodate production.

Design Brief The family of shafts, currently being produced by a manual system, is illustrated in Fig. 10/1. Certain critical features and dimensions (as identified on each shaft) must be maintained; all other dimensions and features are non-critical.

The shaft blanks can be presented in any position and orientation that is convenient and a steady supply will be maintained. After completion the shafts must be deposited on an outgoing conveyor, placed for convenience. Assume that any interlocking and synchronisation between the robot and the machine tools, and between the machine tools themselves, can be readily accomplished.

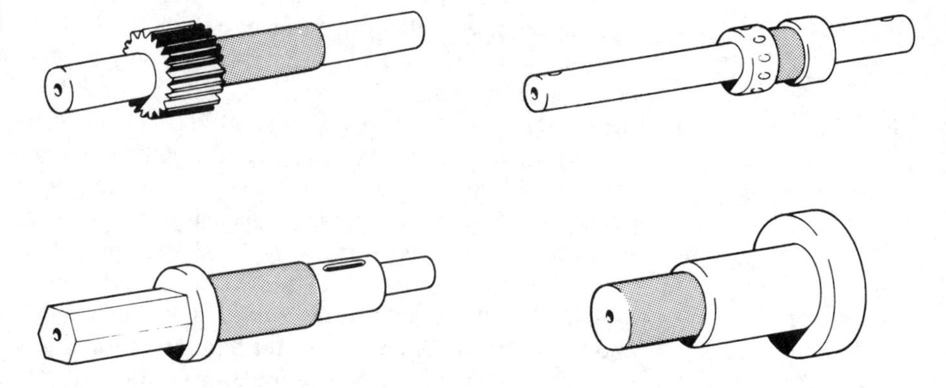

Fig. 10/1 Shafts to be produced by robotised automation

The last operation must be the automatic gauging of the critical dimensions. It can be assumed that an adaptive control system will monitor and adjust machine settings.

Assignment Tasks With reference to the component details of Fig. 10/1:

1 Determine the operations required to finish-machine each shaft axle and formulate a list of machine tools required to form the production cell. Make estimates of the cycle times required to complete each operation and organise the operations into a logical production sequence.
2 With due regard for the true physical size of the machine tools that you specify, the operation sequences determined and your projected cycle times, produce a cell layout for the machine tools assuming the presence of robotised material handling between operations.
3 With due consideration of cell layout, physical robot size and the possibility of alternative robot mounting arrangements, select a suitable industrial robot (or robots) to service production requirements. Compare your specification with the example specifications that appear at the end of this chapter and state which of these, if any, would be suitable. State, giving supporting reasons, the mode of programming that you would suggest for this particular application.

10.1/3 Case Study 3

Application A company produces a range of plastic moulded cases for use in electricity supply meters, gas supply meters, light meters, electronic component cabinets and housings of various other sorts. The plastic moulded cases vary in size, shape and design but each comprise a moulded hollow body onto which is attached a plastic back plate. Some examples of the products are illustrated in Fig. 10/2.

Present System At present the plastic back pieces are individually drilled and screwed onto the moulded cases. Different case designs require different numbers of screws. Because of the variety of designs, drilling of the screw holes and assembly of the back panels are carried out entirely by hand. The number of operators varies with demand and part-time labour is often recruited at peak times. Concern has been expressed that the volume of screws being used is

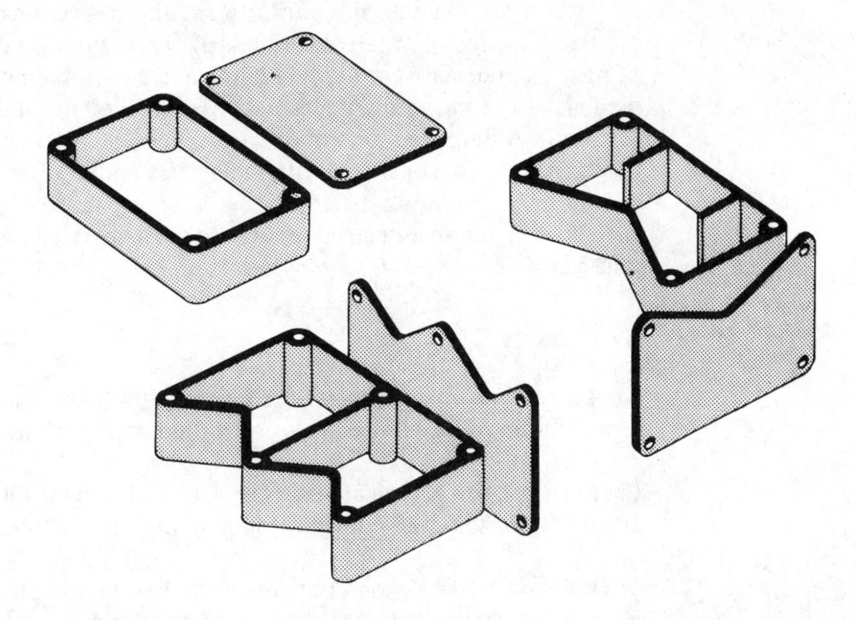

Fig. 10/2 Product examples of moulded plastic cases and screwed lids

becoming excessive, quality of the assembled cases is erratic, and assembly times are variable. Furthermore, feedback from customers has highlighted the following problems which must be addressed:

a) Some applications require the cases to hold liquids, which leak through the screwed joint.
b) Problems of unauthorised access into the cases have arisen by the securing screws being removed.
c) Dissatisfaction has been expressed due to the poor fitting nature of some of the plastic back panels.
d) In applications where the cases have been mounted under conditions of vibration, the cases have become loosened from the back panels.

Manufacturing Brief It has been decided that a study into the design and manufacture of the plastic cases should be carried out. The terms of reference of the study are to:

a) Determine the reasons for the variability in quality and assembly times and the high volume of consumables.
b) Suggest alternative designs, modifications or manufacturing methods that will eliminate the problems highlighted by customer feedback.
c) Suggest the organisation of a manufacturing system that will be flexible and responsive to the needs of producing consistent, high-quality cases under variations in demand.

Assignment Tasks Based on the above historical background and stating any assumptions that you need to make:

1 Write a brief report outlining possible reasons for the variability in quality and assembly times and the high usage of consumables.

2 By considering design modifications, alternative means of fabricating the plastic cases, and alternative methods of production, produce a brief summary of possible alternatives that will eliminate the problems caused by customer feedback. Use the study of alternatives presented in Case Study 1 as a guide.
3 With an emphasis on robotisation, suggest the organisation of a manufacturing system consistent with the needs of point *c* in the terms of reference of the manufacturing brief. Your findings should include details of a specification for an industrial robot, and the associated mode of programming, that can be used within your system.

10.1/4 Case Study 4

Application A manufacturer wishes to evaluate the feasibility of assembling 3-pin, 13-amp electrical plugs using an industrial robot.

Present Design A typical design of a 13-amp plug is illustrated in Fig. 10/3. It consists of some 15 separate component parts as follows:

a) A two-piece plastic body held together by a captive screw.
b) Three brass pins each designed to secure electrical cables by small clamping screws.
c) A fibre cable clamp secured by two small self-tapping screws.
d) A means of securing the fuse.
e) A standard 13-amp fuse.

Assignment Tasks
1 Evaluate the present design and present materials used in the make-up of the 13-amp plug shown in Fig. 10/3. By component re-design and/or use of alternative materials, establish a new design of plug using less than 15 separate component parts and which will lend itself to assembly by automation.
2 Prepare a feasibility report on the use of robotic assembly techniques for this particular product. Your report should consider such points as:

a) Robot type, specification, size, control system and programming capability.
b) Support equipment (hardware) required to support the assembly process.
c) How the component parts should be oriented, fed and presented to the process.
d) Any jigs or fixtures that may be required.
e) The sequence of the assembly process.
f) Any sensing devices that may be required.
g) The type, number and nature of the end effector(s) to be used.

3 For *one* of the components within your re-designed assembly, design an end effector gripper capable of performing the assembly of that component.

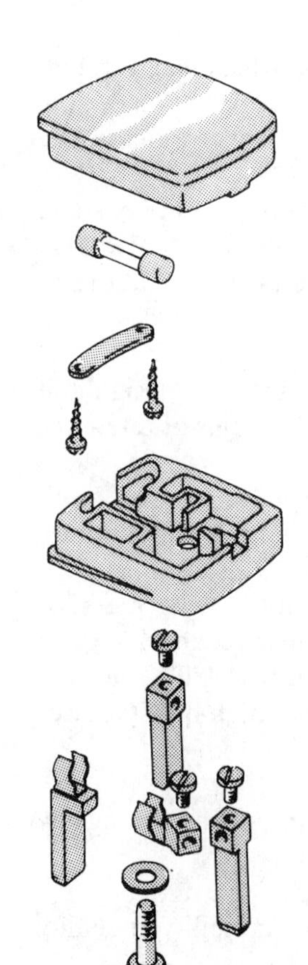

Fig. 10/3 Component parts of a 3-pin 13-amp electric plug

10.1/5 Case Study 5

Application A company produces powder cement, in 50 kg paper sacks, for distribution to wholesalers, builders merchants and DIY retail outlets. It wishes to increase its output to meet rising orders and introduce a second product size (28 kg) to its range.

Present System The final operations within the process include the automatic filling of heavy brown paper sacks to weights of 50 kg. The sacks are then automatically sealed and placed on a continuously running roller conveyor which transports the sacks to a despatch area. The conveyor runs at a fixed speed and the sacks take up a random position as they are transported. In the despatch area two operators manually lift the sacks from the conveyor and proceed to stack them onto wooden pallets, six high. The pallets rest on the floor. When a pallet has been fully stacked one of the operators uses a fork lift truck to transfer the loaded pallets onto a loading bay which is opposite to, and slightly higher than, the roller conveyor. The present arrangement is illustrated in Fig. 10/4*a*.

Fig. 10/4 Palletising application details

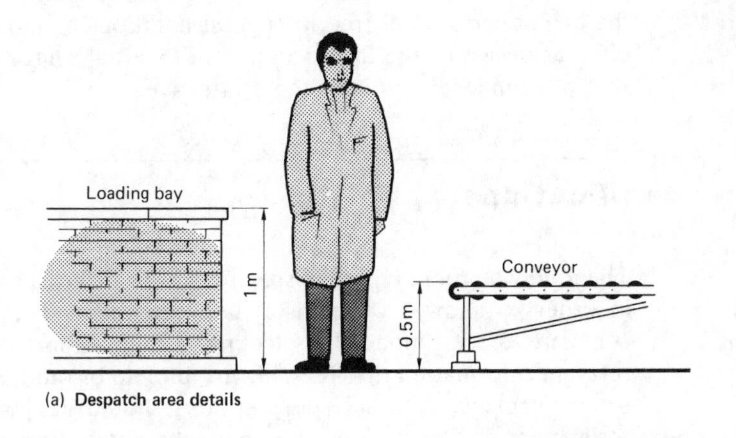

(a) Despatch area details

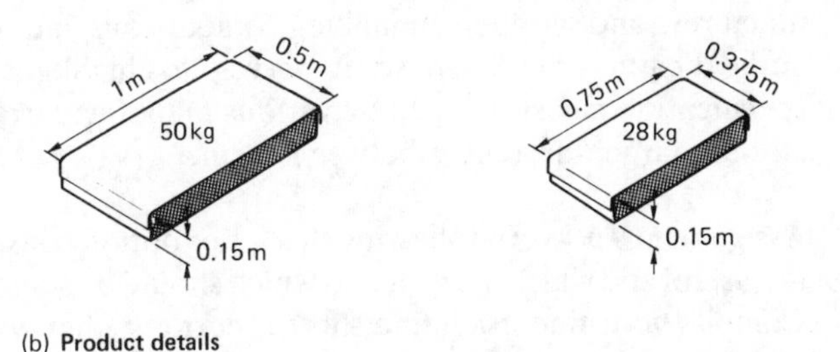

(b) Product details

Assignment Tasks You are asked to investigate the suitability, or otherwise, of robotising the activities in the despatch area.

1 Prepare a draft report justifying the implementation of a robotised palletising system. You may assume that automatic filling of 28 kg sacks can be readily accommodated within the present system, and that the different size sacks will be produced in batches of varying size at short notice. Your report should address such factors as:

a) Comments regarding the company objectives of introducing the second product size and increasing productivity.
b) Effects on the existing workforce in the despatch area.
c) Changes (if any) in workplace layout to accommodate the new system.
d) Modifications or additions (if any) to the conveyor system.
e) The type and specification of industrial robot, control system and programming method, most suitable for the task.
f) The most appropriate type and design of end effector.

2 Assuming a nominal standard pallet size of 1.5 m × 1.0 m, derive efficient arrangement patterns for both 28 kg and 50 kg sacks to make optimum use of pallet storage area. Sack dimensions are indicated in Fig. 10/4*b*. The sacks should be stacked such that the maximum payload placed on the pallets does not exceed 1500 kg. They should be arranged to maximise the number of units per pallet and offer inherent stability and rigidity during handling.
3 For both arrangement patterns produce a flowchart describing the rules to be followed, by the robot, to stack the sacks on the pallets. The aim should be to produce the shortest and most efficient 'algorithm' to effectively stack the components. Feel free to suggest additions or modifications to the end effector design or the incorporation of external sensor devices, for example that will render the process more effective.

10.2 Robot specifications

The following robot types are specified according to a common set of features in order to allow a comparison between them. They are not necessarily complete specifications. In selecting a robot from the specifications, any relevant information that is either incomplete or omitted should be assumed for convenience. All such assumptions should be justified as part of the assignment documentation. Alternatively, obtain literature from robot manufacturers and suppliers in making comparisons and selections. In addition to authenticating the exercise it serves to highlight the large amount of specification measures used by various manufacturers to specify their robots, and the large differences between them.

Assignment Task In following this latter option consider which specifications are useful and which are not, and which should be standard and which optional. Compile your findings into a short report together with your thoughts on the possibility of laying down standards for robot specifications. Your thoughts should address issues such as:

a) The desirability of such standards.
b) The contributors to, and administrators of, such standards.
c) Any problems that would be encountered in such a scheme.
d) How such a standard would be implemented.
e) How such standards would be maintained, updated, etc.
f) How potential manufacturers and suppliers would be identified.

Robot 1

CONFIGURATION	:	Scara
NUMBER OF AXES	:	4
DRIVE SYSTEM	:	DC stepper motor & Pneumatic Z-axis (Vertical)
CONTROL SYSTEM	:	Continuous path
PAYLOAD CAPACITY	:	2kg
POSITIONING ACCURACY	:	±0.05mm
REPEATABILITY	:	±0.25mm
PROGRAMMING METHOD	:	Drive-through, Coordinate entry Full editing facilities
MEMORY CAPACITY	:	999 Steps
MEASURING SYSTEM	:	–
I/O CAPABILITY	:	Digital 8-input, 8-output RS232 communication
BACKING STORE	:	External microcomputer
MOVEMENT CAPABILITY	:	See Diagram H (Height) – 250mm manual adjustment X (Shoulder) – 200° Y (Elbow) – 180° Z (Vertical) – 60mm Reach – 410mm
SPECIAL FEATURES	:	Manual adjustment to set Z-axis

142 202 66
148 – 308
60
510
125 sqr

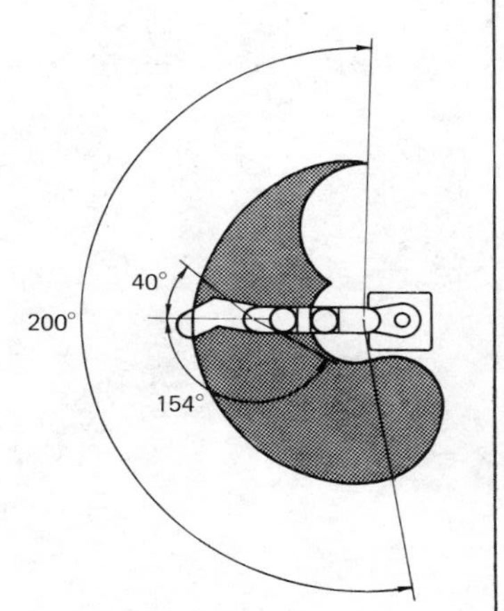

Robot 2

CONFIGURATION	:	Articulated – Closed kinematic
NUMBER OF AXES	:	3 + 3 wrist motions
DRIVE SYSTEM	:	DC PWM servo motor
CONTROL SYSTEM	:	Continuous path with circular interpolation
PAYLOAD CAPACITY	:	60kg
POSITIONING ACCURACY	:	±1mm
REPEATABILITY	:	±0.3mm
PROGRAMMING METHOD	:	Drive-through, Coordinate Entry, Teach pendant
MEMORY CAPACITY	:	32K RAM (2000 points, 1000 instructions)
MEASURING SYSTEM	:	Incremental digital positioning control
I/O CAPABILITY	:	Digital 48-input, 16-output RS232 communication Magnetic cassette
MOVEMENT CAPABILITY	:	See Diagram H (Height) – 2000m X (Shoulder) – 85° Y (Elbow) – 65° Z (Waist) – 300° Reach – 1800mm Wrist – 360° – 360° 200°
SPECIAL FEATURES	:	Tool coordinates Coordinate transformation Dry run operation Inbuilt diagnostics

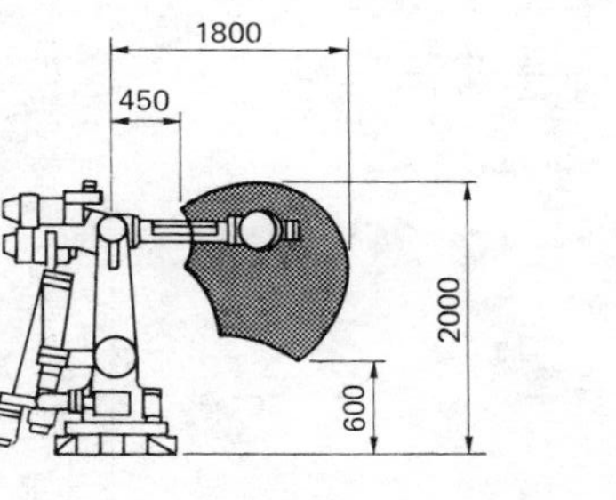

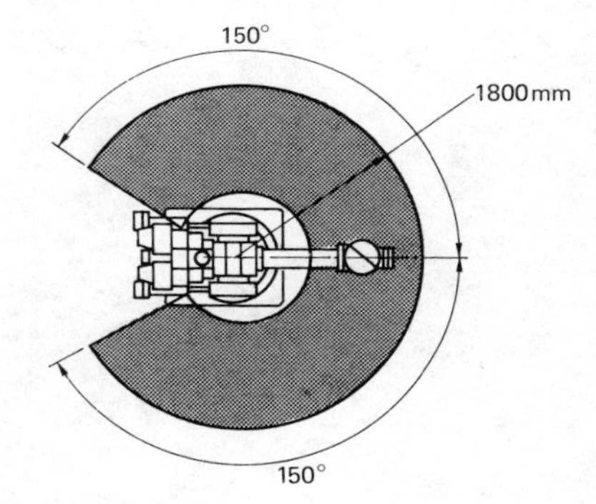

Robot 3

CONFIGURATION	:	Articulated – Revolute
NUMBER OF AXES	:	3 + 1 wrist motion
DRIVE SYSTEM	:	DC servo motor
CONTROL SYSTEM	:	Point to point, Continuous path
PAYLOAD CAPACITY	:	3 kg
POSITIONING ACCURACY	:	±0.3 mm
REPEATABILITY	:	±0.1 mm
PROGRAMMING METHOD	:	Drive-through, Coordinate entry, Off-line Full editing facilities High-level programming language
MEMORY CAPACITY	:	16K RAM
MEASURING SYSTEM	:	Incremental encoder
I/O CAPABILITY	:	Digital 16-input, 16-output RS232 communications
BACKING STORE	:	Floppy disc
MOVEMENT CAPABILITY	:	See Diagram Waist – 320° Shoulder – 220° Elbow – 120° Reach – 1250 mm
SPECIAL FEATURES	:	High-level programming language Inbuilt diagnostics CRT or TTY terminals Floppy disc program storage Right-hand or left-hand configuration

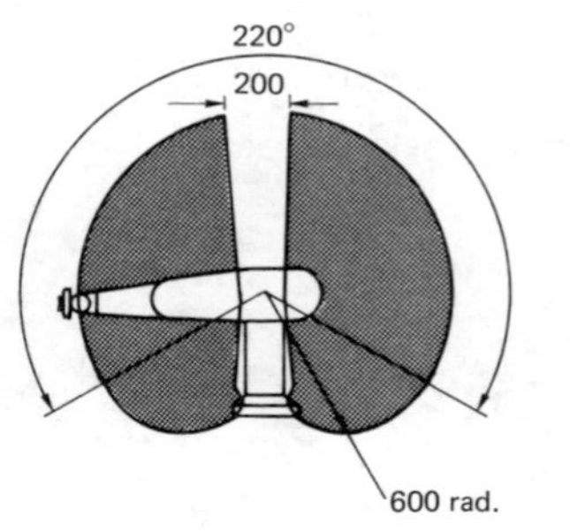

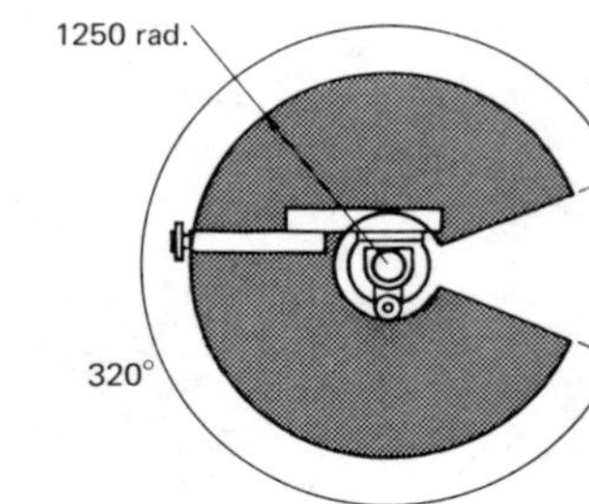
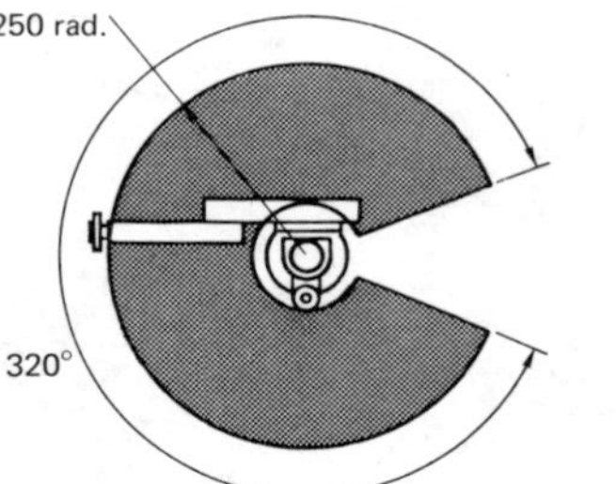

Robot 4

CONFIGURATION	:	Pendular
NUMBER OF AXES	:	3 + 3 wrist motions
DRIVE SYSTEM	:	DC servo motor & Hydraulic servo
CONTROL SYSTEM	:	Point to point
PAYLOAD CAPACITY	:	90 kg
POSITIONING ACCURACY	:	±1.5 mm
REPEATABILITY	:	±1 mm
PROGRAMMING METHOD	:	Drive-through, Coordinate entry Full editing facilities
MEMORY CAPACITY	:	1500 instructions
MEASURING SYSTEM	:	Incremental encoder and grating
I/O CAPABILITY	:	Digital 8-input, 8-output Analog, 4 channels, ±12 V
BACKING STORE	:	Floppy disc
MOVEMENT CAPABILITY	:	See Diagram
SPECIAL FEATURES	:	Programming language available End effector service connections Joystick control Question/answer programming

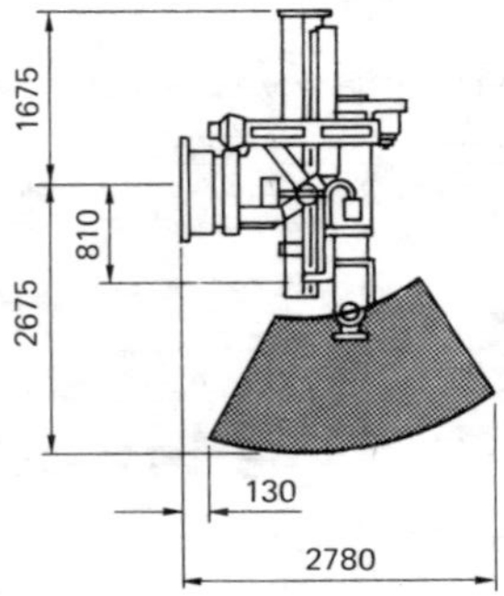

1675
810
2675
130
2780

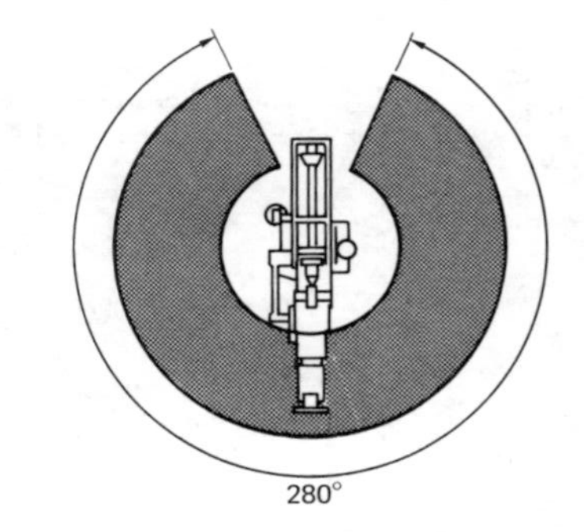

Robot 5

CONFIGURATION	:	Cylindrical
NUMBER OF AXES	:	3
DRIVE SYSTEM	:	DC servo motor
CONTROL SYSTEM	:	Point to point
PAYLOAD CAPACITY	:	5kg
POSITIONING ACCURACY	:	±0.2mm
REPEATABILITY	:	±0.05mm
PROGRAMMING METHOD	:	Drive-through, Coordinate entry, Teach pendant Full editing facilities
MEMORY CAPACITY	:	300 positions
MEASURING SYSTEM	:	High-resolution encoder
I/O CAPABILITY	:	Digital 6-input, 6-output RS232 communication
BACKING STORE	:	Magnetic tape
MOVEMENT CAPABILITY	:	See Diagram Linear horizontal – 500mm Linear vertical – 280mm Rotary swivel – 270°
SPECIAL FEATURES	:	

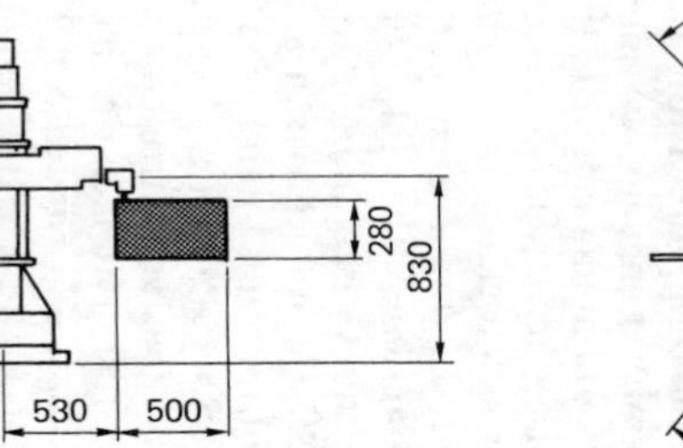

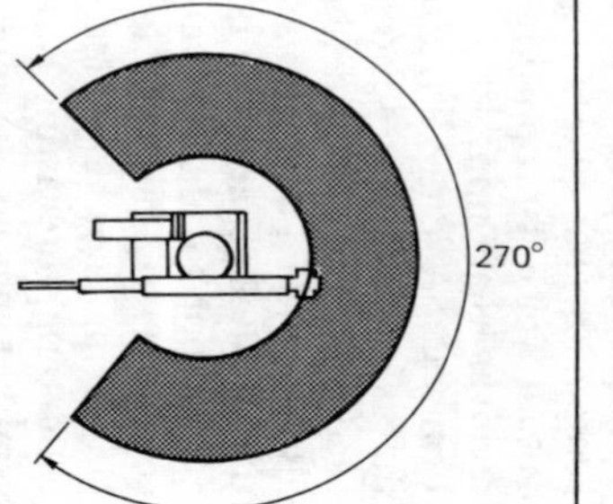

Robot 6

CONFIGURATION	:	Polar
NUMBER OF AXES	:	3
DRIVE SYSTEM	:	DC servo motor
CONTROL SYSTEM	:	Point to point
PAYLOAD CAPACITY	:	30kg
POSITIONING ACCURACY	:	±1mm
REPEATABILITY	:	±0.5mm
PROGRAMMING METHOD	:	Drive-through, Coordinate entry Full editing facilities
MEMORY CAPACITY	:	5000 steps
MEASURING SYSTEM	:	Absolute encoder
I/O CAPABILITY	:	Digital 15-input, 15-output RS232 communication
BACKING STORE	:	
MOVEMENT CAPABILITY	:	See Diagram Waist – 320° Shoulder – 20° Arm extension – 1800mm
SPECIAL FEATURES	:	Any mounting position

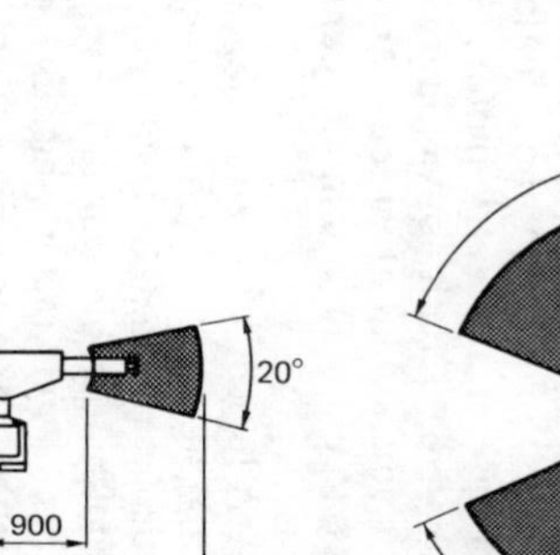

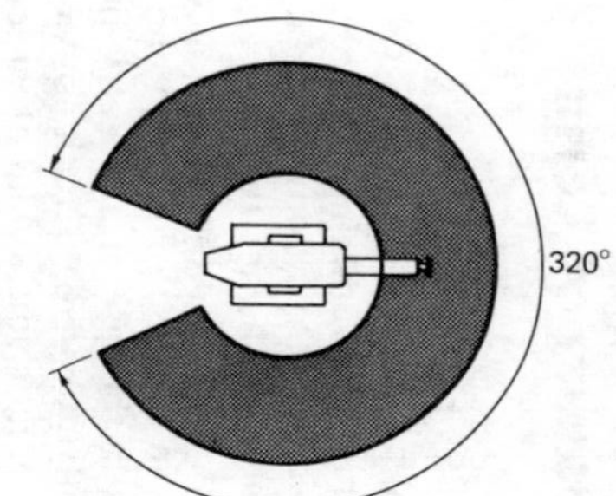

Glossary of Terms

Absolute Coordinates All dimensions are referenced from a fixed point.

Absolute Programming All axis movements are specified in relation to a fixed (datum) point.

Actuator A device that converts a power medium and produces physical movement. May be linear (as in the case of a piston and cylinder) or rotary (as in the case of an electric motor).

Adaptive Control Continuous monitoring and automatic adjustment of various parameters during a process, to maintain optimum operating conditions.

Algorithm A logical, ordered strategy designed to indicate the steps to be carried out in the solution of a particular problem or task, or to show that a solution to that problem or task does not exist.

Alphanumeric A character which is either a letter of the alphabet or a numerical digit.

Analog Quantity A quantity that can vary continuously with time.

Analog-to-Digital Convertor (ADC) Electronic component that gives a digital number as an output which is proportional to a varying voltage it receives as an input.

Articulated A jointed structure, capable of motion, that is made up of a series of rigid links. When used to describe a particular robot structure it may also be termed *anthropomorphic* or *revolute*.

Artificial Intelligence The characteristic of automation to adapt to various (unknown) conditions or changes in its environment, employing knowledge, learning, logic, reasoning, estimating, forecasting, strategy and experience.

Assembly Language A mnemonic-based programming language that addresses the functions of a microprocessor directly. So called because a program called an *assembler* converts (assembles) the program source code into machine language.

Automation The ability of machines to operate without direct human intervention.

ASCII Code American Standard Code for Information Interchange. Standard coding system of 8-bit data for alphabetic, numeric, punctuation and special control characters for data representation communication and storage.

Automatically Guided Vehicle (AGV) A self-contained, free-ranging, driverless vehicle whose route can be modified by remote or on-board computers. Used to load, unload and transport various cargoes within an overall materials handling system.

Axis Direction of motion of a particular machine or robot movement. May be either linear or rotary and may be identified by a letter of the English or Greek alphabet.

Backing Store Mass storage devices or media for the permanent storage of information, programs and data. Usually punched tape, magnetic tape or magnetic disc. Characterised by relatively long access times.

Backlash The term given to describe the amount of lost motion in a mechanical transmission device.

Baud Rate Bits per second. An expression of data transmission speed between two computer devices connected for serial transmission. As a rule of thumb, (baud rate/10) gives the transmission speed in characters per second.

Binary A numbering system employing two digits 0 and 1, which forms the basis of digital computer operation.

Binary Coded Decimal (BCD) A special form of binary code in which each decimal digit is expressed as a 4-bit binary pattern. Large or small decimal numbers can then be translated by coding each decimal digit forming the number, in turn.

Bit Contraction of BInary digiT. A single digit in a binary number. Each bit in a binary number has a weighting dependent on its position. The rightmost bit is termed the Least Significant Bit and the leftmost bit is termed the Most Significant Bit.

Black Box A concept of visualising the operation of any system by considering just the input(s) to and the output(s) from that system. Since it is not necessary to know how the system works (just how it affects the inputs to produce the outputs), the system is known as a black box.

Block Diagram A simplified means of representing the overall operation of a system on paper. The elements of the system are represented by symbols. The interconnection, inputs and outputs of the various elements are represented by straight lines. The direction of information flow or movement is indicated by arrowheads on the interconnection lines.

Branching A programming mechanism that transfers execution of the program to an earlier or later part of the program. May be conditional, where some condition must be satisfied before the branch takes place, or unconditional where the program always branches.

Buffer A temporary holding store between two stages of a process which are operating at different speeds. In a robot control unit it is an area of random access memory through which data passes when being transferred from one device to another.

Byte A group of 8 bits. Each byte is capable of decoding one character of information. Computer memory capacity is measured by the number of characters (bytes) it can store. The term *kilobyte* or K represents 1024 bytes and is used as an expression of memory size.

CAD/CAM Computer Aided Design/Computer Aided Manufacture. General terms used to describe the application of computer technology to the engineering functions of producing engineering designs and components.

Cartesian Coordinates A system of coordinates where features are located by dimensions at right angles to each other. Also known as *rectangular coordinates*.

Cartesian Configuration A robot structure which allows three mutually perpendicular motions and movement description by rectangular coordinates.

Character A single unit of information. May be a digit, letter of the alphabet, punctuation symbol, graphics symbol or control symbol that can be represented as one byte of information.

Charge Coupled Device (CCD) Electronic device in which data is stored by electrical charge.

Checksum An error checking device for transmitted data. Each byte of data is manipulated arithmetically to produce a coded check digit that is transmitted along with the data. The device receiving the data also decodes the check digit. Any difference between the two values indicates an error in transmission.

Closed Loop Control System of machine control in which various output conditions (e.g. slide position and velocity) are continuously monitored and compared with the input (command) signal. The monitoring and comparing operations are termed *feedback*.

Compliance Quality of a device or material to 'give' or adjust to its service conditions. Opposite to rigidity.

Computer Integrated Manufacture (CIM) A total design concept of a computer controlled manufacturing environment. An essential feature of CIM is that a single source of manufacturing data (manufacturing database) can be accessed and used by all the different functions of the organisation.

Computer Numerical Control (CNC) A system of controlling machine tools by coded, alphanumeric instructions contained within a dedicated, stored-program computer control unit.

Conditional Branch A programming construct enabling the flow of a sequence program to be modified according to a stated condition. In digital systems only two-state conditions are permitted, e.g. pass/fail, true/false or yes/no, and the branch may be forwards or backwards in the sequence program.

Configuration Term given to describe the physical design and geometrical structure of an industrial robot.

Contact Sensor A sensing device activated by physical contact.

Continuous Path Control A type of control system in which all points along a path of motion are specified or generated.

Control Engineering An engineering discipline dealing with the design and operation of automatically controlled systems.

Conversion Time An expression of the speed at which an A-to-D convertor can convert an analog voltage into its digital representation.

Cyclic Redundancy Check An error checking device on data transmitted via electronic means.

Cyclo Drive A mechanical speed reducer employing rotating elements having teeth of cycloidal form.

Cylindrical Coordinates A system of coordinates whereby any point is specified by an angular dimension and two linear dimensions, within a cylindrical work space.

Cylindrical Robot A physical robot configuration constrained to move within a cylindrical working envelope.

Cylinder Fluid power element in which a piston (sliding within a cylinder) under fluid pressure produces linear motion. May be either double acting in which movement in both directions is under power, or single acting in which powered motion is achieved in one direction only.

Damping Resistance to motion introduced to reduce undesirable oscillations and counter the effects of excessive overshoot or undershoot. The absorption of energy.

Data Direction Register Memory location within a dedicated input/output integrated circuit, that determines the status (either input or output) of data signal connections.

Data Register Memory location within a dedicated input/output integrated circuit, that latches input or output signals on data signal connections.

Datum Fixed reference point about which movements or measurements are made.

Dead Zone Any area within the working envelope of an industrial robot that cannot be reached by the end of the robot arm.

Dedicated Computer A computer system dedicated to carrying out a fixed, clearly defined task.

Degree of Freedom One of six motions sufficient to describe the position and orientation of a body in space.

Degree of Movement A movement provided by a linear or rotary joint within a robot structure.

Digital Quantity A quantity that can be represented by one of two mutually exclusive states. Since it does not matter what actual values represent the two states, they are referred to as 1 and 0.

Digital Computer A computer system that operates using digital voltage levels.

Digital-to-Analog Convertor (DAC) Electronic component that gives, as an output, a varying voltage proportional to a digital number it receives as input.

Digitise The process of converting data into a digital form.

Dimensional Tolerance The amount of permitted error allowed (either above and/or below) from a stated dimension.

Direct Drive Drive system powered directly by the actuator and not involving intermediate transmission elements.

Direct Numerical Control The direct control of one or a number of CNC machine tools, robots or other advanced manufacturing technology, by a host computer. The downloading of sequence programs from a host computer directly into the memory of a robot control system.

Editing Facility whereby the machine operator can modify a stored program and/or information registers, in the memory of a robot control unit, without altering the original program medium from which the stored program was read.

Encoder A type of transducer commonly used to convert angular or linear movement into digital data as a means of determining absolute or incremental position.

End Effector General term used to describe any end-of-arm tooling attached to the end of a robot manipulator in order to enable it to carry out a specific task. End effectors may be tools, sensors or grippers and often incorporate two or three wrist movements.

EPROM Erasable Programmable Read Only Memory. A programmable memory chip whose contents remain intact even when the power source is removed (non-volatile). The memory chips can be erased by exposing them to ultra-violet light for a short period of time, and then re-programmed.

Executive Master control software.

External Sensor A sensor for determining feedback information in the environment external to the robot.

Fail-safe Failure of a device or system without damage or danger, as provided for by built-in design features.

Feedback A feature of closed loop control systems whereby an electrical signal, proportional to a monitored quantity (axis position or velocity), is compared with the command signal. The result of the comparison is an indication of the error between the commanded and the actual condition.

Firmware A general term used to describe programs (software) held on a memory chip (hardware).

Fixture A device for locating and holding one or more components during production operations.

Flexible Manufacturing System A means of organising production using high levels of automated equipment and computer control to achieve largely unmanned manufacture.

Flowchart A form of graphical shorthand, using accepted symbols, used to plan, document and communicate action sequences.

Fluid Power An actuation system using fluid pressure to transmit force and motion. Fluid media are usually compressed oil (hydraulics) or compressed air (pneumatics).

Gain Design function of a control system that is a measure of the amplification required between input and output signals to balance the behaviour of the system.

Generation A distinct phase of advancement in the development of a particular technology, as in third-generation robots or fifth-generation computers.

Gripper An end effector that uses mechanical, vacuum or magnetic means to lift or manipulate cargoes. May be termed *hard* if they are not capable of assuming the shape of a cargo, or *soft* if they can take up the shape of a number of different cargoes.

Handshaking A system used between two interconnected devices which operate at different speeds, to ensure that all data transmitted by the transmitting device is received by the receiving device. Control signals signalling 'ready to receive' and/or 'ready to transmit' must be present to initiate data transfer.

Hardware A general term used to describe all physical components of a robot or computer system.

Hard Automation Automation that is purpose-designed and built to carry out a specific task, the configuration of which cannot easily be changed.

Harmonic Drive A single-stage mechanical speed reduction device offering reduction ratios in excess of 320:1.

Heuristic Learning by discovery, experience, reason and experiment.

Hidden Costs In a manufacturing environment, those costs that can significantly affect the overall costs of production but which cannot easily be accurately determined.

High-level Language Computer programming languages that use meaningful words and phrases which are relatively easy to learn and quick to program.

Host Computer A master control computer that directs the operation of one or more associated peripheral elements. Any computer from which a robot or machine tool accepts instructions or downloaded program data.

Hunting Feature of a closed loop control system whereby a moving axis slide oscillates about its commanded destination position. Caused by a poorly designed closed loop control system trying to compensate for the effects of excessive overshoot and/or undershoot when negotiating a target position.

Hydraulic Systems or system components that utilise fluid in the form of a liquid to transmit power. Hydraulic power may also be referred to as *fluid power*.

Incremental Coordinates All dimensions are referenced from a previously dimensioned point rather than a fixed reference point.

Incremental Programming All axis movements are specified in relation to the last point visited rather than from a fixed (absolute) point.

Inductosyn A position measuring transducer working on a similar principle to the synchro resolver. Often used for fine measurements.

Industrial Robot A robot used in a manufacturing or production environment for the purpose of performing one or more industrial tasks.

Input/Output A means of describing facilities and signals used when external devices communicate with computerised systems.

Instability See *Hunting*.

Interface A collection of electronics designed to make output signals from one device compatible with the input signal requirements of another device.

Internal Sensor A sensor used for measuring dynamic quantities, such as position, velocity, acceleration, etc., internal to the robot.

Interpolation The joining up of programmed points to generate a smooth path by computation. If the segments joining the points are straight lines, the process is called *linear interpolation*. If the segments joining the points are arcs of circles, the process is called *circular interpolation*, and if the segments joining the points are arcs of a parabola, the process is called *parabolic interpolation*.

Interrupt (Service Routine) An interruption of the operation of a microprocessor due to an external device requiring service. A routine that enables a microprocessor to break off from the task it is engaged upon to service an external request. After servicing the request, the microprocessor will resume the task at the point at which it was interrupted.

Intrusion Monitoring A safety system designed to alarm out or initiate a controlled shutdown of a robotised operation if unauthorised persons (intruders) enter a designated no-go area.

ISO Code System International Standards Organisation standard coding system for the representation of characters on 8-track punched tape. Track 8 is used for parity checking.

I/O (Chip) Abbreviation for input/output. An integrated circuit dedicated to handling data signals for either input to, or output from, a digital computer system. May also be termed a PIO (Programmable Input/Output) chip, a VIA (Versatile Interface Adaptor) chip, or simply an I/O port.

Jogging Method of manually controlling robot axis movement. The depression of a jog button moves the selected axis by a fixed amount of movement. The amount of movement per depression, the direction of movement and the axis are all selectable by the operator. May also be referred to as *inching*.

Knowledge Engineer A computer scientist able to extract relevant information, knowledge and experience (usually from experts), and represent it in a computerised form that can be readily accessed by non-experts.

Lead Time The amount of time that elapses between the receipt of an order and delivery of that order.

Locomotion Providing movement to the whole of a robot structure to increase its versatility, flexibility and working envelope.

Locus The shape of a path traced out by a moving point constrained to move under certain conditions. (Plural: *loci*.)

Loop A repetition function of a computer or robot sequence program language. Allows a defined sequence of steps to be repeated a desired number of times. A means of causing the flow of a program to be transferred to a previous point in the program.

Low-level Language Computer programming language that uses instructions closely related to the fundamental instruction set of the microprocessor.

Machine Control Unit (MCU) Term given to describe the physical components of a robot control system. Includes the control console, VDU, backing store, input keyboard and teach pendant.

Machine Intelligence The capability of a machine to sense conditions and changes in its environment and modify its actions accordingly.

Machining Cell An organisation of a small number of machine tools arranged to manufacture a defined set of components.

Macro A facility of a programming language that allows a specified number of commands, in the form of a sub-program, to be executed by a single program call. May include the facility to pass parameters to the sub-program.

Manipulator The physical element of an industrial robot system that provides main movement and load-carrying capability.

Manual Data Input (MDI) The manual 'keying in' of sequence programs to the memory of a robot control unit via the console keyboard or teach pendant.

Mark/Space Ratio The ratio of the applied voltage time to the non-applied voltage time in a pulse-width-modulation motor control system.

Matrix An array of rows and columns.

Memory Any device capable of storing information for retrieval at a later time. See also RAM, ROM and EPROM.

Micro-electronics The miniaturisation of electronic circuits and components.

Microprocessor A programmable integrated circuit (chip) that forms the heart of many general-purpose and dedicated computer control systems.

Mnemonic A shortened form of a word or phrase so constructed as to indicate the meaning or function of the word or phrase it represents. Three-letter mnemonics form the instruction sets of most low-level assembly languages.

Modular Programming A technique of programming which breaks the total program down into short, stand-alone modules (sub-programs) that carry out an easily identifiable task or process. The modules can be individually tested and used as building blocks for use in other completely different programs.

Moiré Fringe A repeating interference pattern of light and dark alternating bands produced as a result of two optical gratings moving relative to each other. There is a mathematical relationship between the number of lines on the gratings and the movement of the fringes for a given movement of the gratings.

Moment of a Force The turning effect of a force acting about a certain point. Quantified as Force × Perpendicular Distance about the point at which it acts.

Multigripper An end effector that incorporates a number of different end effector elements within a single unit. Reduces the need to carry out an end effector change and/or enables a number of different components or tasks to be serviced using a single end-of-arm tooling unit.

Negative Feedback A control system concept. The comparing of a monitored output signal with a commanded input signal by subtraction. A result of zero indicates that the actual value coincides exactly with the commanded (target) value. A non-zero result indicates that the target value has not yet been reached or has been exceeded and corrective action is required.

Noise A term used to describe electrical interference.

Open Loop Control A control system in which there is no feedback. Position and velocity are determined by inbuilt features of the driving mechanism.

Operating Program The master control program that directs the action of a robot control system. Held in non-volatile, read only memory. It is provided by the control system designers and cannot be modified by the end user. Sometimes called the *executive*.

Overhead An expense item that does not contribute directly to the manufacture of a component or service. Such costs have to be borne by sales of the component or service.

Palletising The process of ordering and placing items according to a set pattern or arrangement, usually on pallets.

Parallel Processing Computer processing tasks carried out concurrently by executing different program instructions at the same time. Requires specially designed computer hardware and software.

Parameter A numerical value that once defined remains constant, but may subsequently be re-defined to a different value for use under the same circumstances.

Parity Check An error checking device for binary transmitted data. Operates by determining whether the number of bits set to binary 1 (in each byte of transmitted data) is odd or even. Depending on the result, and whether odd or even parity is specified, a reserved bit (within the same byte) is either set or re-set to maintain the parity condition. Commonly applied to punched tape.

Parts Presentation The alignment, positioning, orientation and workholding of component parts prior to undergoing an automated process.

Pattern Recognition The computerised recognition of digital images.

Payback Period A period of time during which the combined benefits derived from the installation and operation of a robot system are deemed to have covered the costs of doing so.

Payload Capacity The maximum weight or mass that can be handled satisfactorily by an industrial robot under normal and continuous operation.

Peripheral General term used to describe a physical piece of equipment used, in conjunction with a computer system, for supporting functions. Common peripherals include disc drives and printers.

Permit to Work A safety system whereby authorised documentation is required before persons are allowed access to a robotised work area.

Photocell An electronic component (transducer) that gives an output voltage dependent on the amount of light falling on it. Some photocells give a constant voltage once a particular light threshold has been reached, whilst others give an analog voltage proportional to the light intensity falling on the photocell.

Pitch, Roll and Yaw Terms given to describe the movements of a robot wrist or end effector.

Pixel Picture element. The smallest unit of resolution of a computer-generated image.

Planetary Roller Leadscrew A leadscrew system in which sliding friction is replaced by rolling friction. Threaded rollers revolve around a threaded leadscrew in a manner similar to the way planets revolve around the sun.

Point-to-Point Control A control system type in which movements are specified as target points in space. The intermediate motion path between points may or may not be in a straight line.

Polar Coordinates A system of coordinates in which points are described by a length from a pre-defined datum point, and an angle from a specified plane.

Polling The sequential testing of an input port to determine the status of incoming signals.

Port A data input/output point on a computerised system. An orifice providing inlet or exhaust in a hydraulic component.

Positional Measuring System Measuring system designed to monitor the position or movement of a controlled axis, to produce feedback in a closed loop control system.

Preprocessing Auxiliary computer processing operation carried out to ease the workload on the main processor.

Printed Circuit Board Electronic circuits produced by discrete components on self-contained boards or cards. May allow the simple replacement of damaged or faulty circuitry and a simple means of updating or enhancing control system facilities.

Prismatic Joint A robot arm movement provided by a linear sliding motion.

Program A sequence of instructions conforming to certain rules of syntax that can be executed by a computer or microprocessor system to perform a particular task.

Program Proving Technique(s) of verifying the safe and correct operation of robot sequence programs.

Proximity Sensor A device which senses that an object is within a short distance of the sensing device.

Pulse Width Modulation (PWM) A technique of controlling DC motors by controlling the time that signal voltage is applied.

Punched Tape A storage medium for the permanent storage, and loading, of data and sequence programs. Punched tape for CNC applications is 25 mm wide and holds 10 characters per 25 mm length. Common tape materials include paper, plastic and polyester laminates.

Random Access Memory (RAM) Computer memory that can be 'written to' and 'read from' (by the computer control system) with equal ease and speed. The contents of RAM are usually lost when the power source is removed (the contents are said to be volatile). Memory contents can be retained by supplying power by battery back-up when the main power source is removed. Supplied as memory chips in units of 1K capacity. Used for holding user-supplied programs.

Read Only Memory (ROM) Computer memory that can only be 'read' by the computer control system. 'Writing to' ROM has no effect. The contents of ROM are permanent and remain intact even when the power source is removed (the contents are said to be non-volatile). The contents of ROM are determined by the application and, once programmed, cannot be modified or re-programmed. Used for holding main control system software.

Real Time Information processing or transmission which occurs in such a way that it can influence the behaviour of a process, whilst that process is operating normally.

Recirculating Ball Leadscrew A leadscrew system in which sliding friction is replaced by rolling friction. Hardened ball bearings re-circulate around a ground form leadscrew.

Register A random access memory location that is internal to a microprocessor. The name given to a single random access memory location used for a specific purpose.

Remote Centre Compliance End effector designs originated for assembly operations. An unpowered, mechanical means of compensating for lateral positioning error and angular misalignment.

Repeatability The closeness of repeated positional movement, under the same conditions and movement specification.

Resolution The smallest increment of a system that can be detected, measured or acted upon.

Resolver A feedback transducer that converts rotary motion into an analog voltage to represent angular position.

Revolute Joint A robot arm movement provided by a rotary motion.

Roll See *Pitch, Roll and Yaw*.

RS232 Interface A collection of electronics that arranges the correct protocol for serial data transmission according to EIA standard RS232C. Such interfaces may be designed for one-way or two-way communication.

Sampling Data acquisition carried out at predetermined time intervals.

Sensor A device capable of responding to an environmental change and reporting that change by means of an output signal. *Internal sensors* are located within the structure of an industrial robot to sense manipulator position, velocity and acceleration. *External sensors* are provided to feed back information concerning the immediate working environment.

Service Unit Element of a fluid power system used for cleaning and conditioning the fluid prior to it entering the circuit.

Servomechanism An automatic control system in which position is the controlled quantity. Incorporates feedback and power amplification to enable an output quantity to follow an input (command) signal without error. Used in industrial robots for axis positioning.

Servomotor A motor comprising integral elements and providing feedback signals for use in automatic control systems.

Sequence Program A taught user program originated to describe the movements of an industrial program, usually to carry out a specific task.

Shaft Encoder A transducer in which the (analog) angular position of a rotating shaft is converted into (digital) coded form using a digitally coded disc.

Single Step A mode of operation of an industrial robot running a sequence program. Individual instructions of the sequence program are executed one at a time upon the depression of a key on the console keyboard or teach pendant. After a block has been executed the robot stops and awaits further commands.

Software General term used to describe computer programs and the media on which they are stored.

Soft Automation Automation that operates under the control of a computer program (software) which can easily be modified or adapted.

Solid State Electronic. No moving parts.

Spool Valve Fluid power component used for controlling the pressure, flow and direction of compressed fluid within a fluid power circuit.

Stack A data/programming structure in a microprocessor within which stored data must be retrieved in the reverse order to which it was stored. Said to employ a LIFO (Last In First Out) structure.

Steady State The condition of a control system when equilibrium of movement has been reached.

Stepper Motor A digital electric motor used in open loop robot control systems for axis movement. The rotor (connected to the axis leadscrew) moves in small, fixed angular steps (in either direction) on receipt of digital pulses from the control system. Rotational speed (hence feedrate) depends on the frequency of the applied pulses.

Subroutine or Sub-program A separately defined part of a computer or sequence program. Used to simplify and shorten programs by defining commonly used sequences once only, and called and executed as required by the main program.

Synchro-resolver An electro-magnetic position transducer whose output voltage depends on the angular position of its rotor.

Tactile Sensor A sensing device that must touch or make physical contact with an object to sense it.

Teach Pendant A remote control keypad device used for programming or manipulating the robot at some distance from the main control system console.

Template Matching Method of shape or pattern recognition by comparing image data against existing stored pattern data.

Tool Coordinates A technique of transforming datum positions and axis movements so that they correspond to the attitude of the end effector.

Track A path along which coded information may be stored on backing store medium. On punched or magnetic tape, a track is a path running the length of the tape and parallel with its edge. On a magnetic disc, a track is a circular path concentric with the driving hub. Sometimes called a *channel*.

Transducer A device that converts energy in one form into energy in another form, in such a way that the output is a known function of the input.

Transformation Mathematical conversion of coordinate positions.

Translation Linear movement or conversion from one quantity to another.

TTL (Levels) Transistor-Transistor Logic. Voltage levels used in digital integrated circuit components and systems.

Turnkey System A system in which a sole supplier takes responsibility for all aspects of its design, installation, commissioning and operation. A system ready for immediate use on purchase and installation.

Unconditional Branch A programming facility within many programming languages that forces the flow of a sequence program to be altered. Branches may be forwards or backwards in the sequence program. May be called an *unconditional jump*.

Vacuum Gripper End effector device employing suction cups to grip suitable cargo objects.

Value Analysis An exercise carried out on existing products to determine whether, by re-design, they can be manufactured more easily or cheaply.

Vibratory Bowl Feeder A continuously operating parts orientation and feeding device, usually for small components. The feeder feeds components along a route comprising various orientating devices such that, when the components eventually reach the output point, they are in the correct orientation and attitude. Components that are dis-oriented are deflected back into the bowl to re-enter the route.

Walk-through Programming Technique of programming (akin to teaching by showing) in which the robot is remotely driven to required positional locations. Also called *drive-through programming*.

Water Jet Cutting A cutting process utilising a fine, high-pressure water jet. Used on materials such as plastics, leather, cardboard, etc.

Wind Up The term given to describe lost motion due to elastic torsional compliance of a rotated shaft.

Working Envelope The space surrounding the robot (in all directions) to which the end of the robot arm has access.

Work In Progress (WIP) Stocks of part-finished components awaiting further processing. High work in progress means large amounts of working capital tied up in part-finished stocks.

World Modelling A programming technique that entails 'teaching' the robot comprehensive details about its immediate working environment.

World Coordinates Coordinates referenced about an absolute datum that remains fixed.

Yaw See *Pitch, Roll and Yaw*.

Zero Shift A facility on an industrial robot whereby the axis datum(s) can be shifted to any point within the programmable area of the robot.

Index